Advances in Drug Discovery Techniques

Advances in Drug Discovery Techniques

Edited by

Alan L. Harvey

Strathclyde Institute for Drug Research and Department of Physiology & Pharmacology, University of Strathclyde, Glasgow, UK

JOHN WILEY & SONS
Chichester • New York • Weinheim • Brisbane • Singapore • Toronto

Baffins Lane, Chichester,
West Sussex PO19 1UD, England

National 01243 779777
International (+44) 1243 779777
e-mail (for orders and customer service enquiries): cs-books@wiley.co.uk
Visit our Home Page on http://www.wiley.co.uk
or http://www.wiley.com

Reprinted September 1999

Commissioned in the UK on behalf of John Wiley & Sons, Ltd by Medi-Tech. Publications, Storrington, West Sussex RH20 4HH, UK

Other Wiley Editorial Offices

John Wiley & Sons, Inc., 605 Third Avenue,
New York, NY 10158-0012, USA

Weinheim • Brisbane • Singapore • Toronto

Co-publisher

Interpharm Press, Inc., 1358 Busch Parkway, Buffalo Grove, IL 60089, USA
Telephone: (+ 1) 708 459 8480
Fax: (+ 1) 708 459 6644

Library of Congress Cataloging-in-Publication Data

Advances in drug discovery techniques / edited by Alan L. Harvey.
p. cm.
Includes bibliographical references and index.
ISBN 0–471–97509–5 (hbk) : alk. paper)
1. Pharmacognosy. 2. Drugs–Design. 3. Molecular pharmacology.
I. Harvey, Alan L., 1950–
[DNLM: 1. Drug Design. 2. Drug Screening. 3. Drug Evaluation.
4. Technology, Pharmaceutical. QV 744 A2445 1998]
RS160.A34 1998
615′.19–dc21
DNLM/DLC
for Library of Congress 97–50470
CIP

British Library Cataloguing in Publication Data

A catalogue record for this book is available from the British Library

ISBN 0 471 97509 5

Typeset in 10/12 pt Times from the authors' disks by Keyword Publishing Services Ltd.
Printed and bound in Great Britain by Biddles Ltd, Guildford and King's Lynn
This book is printed on acid-free paper responsibly manufactured from sustainable forestry, in which at least two trees are planted for each one used for paper production.

Contents

Contributors

C. Frank Bennett	ISIS Pharmaceuticals, 2280 Farraday Avenue, Carlsbad, CA 92008, USA
Nicholas M. Dean	ISIS Pharmaceuticals, 2280 Farraday Avenue, Carlsbad, CA 92008, USA
Tamotsu Furumai	Biotechnology Research Center, Toyama Prefectural University, 5180 Kurokawa, Kosugi-machi, Toyama 939-0398, Japan
Kerstin Gunnarson	Biacore AB, Rapsg. 7, Uppsala, Sweden
Alan L. Harvey	Strathclyde Institute for Drug Research and Department of Physiology & Pharmacology, University of Strathclyde, 204 George Street, Glasgow G1 1XW, UK
Diana C. Hill	Xenova Ltd, 240 Bath Road, Slough SL1 4EF, UK
K. Peter Hirth	SUGEN, Inc., 351 Galveston Drive, Redwood City, CA 94063, USA
Iain S. Hunter	Strathclyde Institute for Drug Research and Department of Pharmaceutical Sciences, University of Strathclyde, 204 George Street, Glasgow G1 1XW, UK
Marcel Jaspars	Marine Natural Products Laboratory, Department of Chemistry, Aberdeen University, Aberdeen AB24 3UE, UK
Sanj Kumar	Biacore AB, 2 Meadway Court, Meadway Technology Park, Stevenage, Hertfordshire SG1 2EF, UK

Gerald McMahon	SUGEN, Inc., 351 Galveston Drive, Redwood City, CA 94063, USA
Brett P. Monia	ISIS Pharmaceuticals, 2280 Farraday Avenue, Carlsbad, CA 92008, USA
Ian J. Norrington	Tecan UK Ltd, 18 The High Street, Goring-on-Thames, Reading RG8 8AR, UK
Toshikozu Oki	Biotechnology Research Center, Toyama Prefectural University, 5180 Kurokawa, Kosugi-machi, Toyama 939-0398, Japan
Christopher J. Pazoles	Phytera Inc., 377 Plantation Street, Worcester, MA 01605, USA
Stephen J. Shuttleworth	Biochem Pharma, 275 Boulevard Armand-Frappier, Laval des Rapides, Quebec, Canada H7V 4A7
Scott Siegel	Phytera Inc., 377 Plantation Street, Worcester, MA 01605, USA
Angela M. Stafford	Phytera Ltd, Regent Court, Regent Street, Sheffield S1 4DA, UK
Flora M. Tang	SUGEN, Inc., 351 Galveston Drive, Redwood City, CA 94063, USA
Bohdan Waszkowycz	Proteus Molecular Design Ltd, Beechfield House, Lyme Green Business Park, Macclesfield, Cheshire SK11 0JL, UK
Peter G. Waterman	Strathclyde Institute for Drug Research and Department of Pharmaceutical Sciences, University of Strathclyde, 204 George Street, Glasgow G1 1XW, UK
Li-An Yeh	Phytera Inc., 377 Plantation Street, Worcester, MA 01605, USA

Preface

The pharmaceutical industry has been spending increasing amounts on research and development, but the number of new medicines marketed each year has decreased. Most companies have made improvements to the development process (toxicology, clinical trials, regulatory submission), and attention is now turning to how to improve the process of discovering the leads that can be developed into new medicines.

Approaches to drug discovery are changing rapidly because of the impact of new technologies. The aim of this book is to provide a guide to the various approaches and relevant techniques. The book does not aim to cover all relevant methods in detail; it provides critical examples and information about current developments by leading practitioners.

The context for modern drug discovery is introduced in the first chapter, which outlines the advantages and disadvantages of rational drug design, random screening, and genetic approaches to drug discovery.

The following chapters deal with the continued importance of natural products as sources of novel chemicals that can act as leads to new drugs. Professor Waterman provides a general review on the use of higher plants and on the advances in spectroscopy techniques that allow rapid determination of chemical structure. Dr Jaspars covers recent work involving compounds from marine sources that are particularly promising as anticancer agents, while Dr McMahon and colleagues discuss a case study involving high throughput cellular assays to discover specific enzyme inhibitors. Dr Hill outlines the use of microbial products, and explains how screening chemical processing and information technology need to be optimized to provide the best use of natural products in drug discovery.

Also relating to natural products, Dr Stafford and colleagues show how the range of chemicals from plants can be increased by the use of tissue culture techniques, and Professor Hunter describes how unique chemicals can be obtained by artificial combinations of microbial enzymes from the synthetic pathways leading to the polyketide antibiotics.

Another source of novel chemicals for screening comes from combinatorial chemistry. Dr Shuttleworth introduces the topic and describes the use of solution-phase combinatorial chemistry for optimization of lead compounds.

Rational drug design, or computer-aided drug design, is discussed by Dr Waszkowycz. He summarizes the advances and points out the limitations of attempts at *de novo* drug design.

Some advances in technology of bioassays are covered in the remaining chapters. Dr Kumar describes how surface plasmon resonance can be a sensitive alternative to the use of radioisotopes, and Ian Norrington describes approaches to the automation of assays through the use of robots.

Finally, the potential of antisense oligonucleotides as therapeutic agents is discussed by Dr Bennett and colleagues.

1

Sense, Nonsense and Antisense: Different Approaches to Drug Discovery

Alan L. Harvey

1.1 INTRODUCTION

1.1.1 Drug discovery in context

The process of going from fundamental research to the development of a medicine is shown in Figure 1.1. The discovery parts of this process refer to the choice of therapeutic target, the discovery of the first activity against the therapeutic target, and the selection of the most promising lead compound. Thereafter, the procedures are more usually considered as part of the development process, in which the candidate compound undergoes toxicological testing, pharmacokinetic tests, studies on absorption, distribution, metabolism and elimination, and, of course, clinical trials leading to eventual marketing of the new product. There is increasing concern within the pharmaceutical industry about the costs of drug discovery and drug development, and there are worries about the diminishing return on commercial products and the long term viability of the present industry structure (see James, 1994; Craig and Malek, 1995; Drews and Ryser, 1997). In the search for greater efficiency and economy, both areas of drug discovery and drug development are being examined to determine where potential savings could be made.

In terms of drug discovery, what aspects concern the industry? First, there is a sense of opportunity in that there are many important unmet therapeutic needs where drug treatment either does not exist or is inadequate. A prime example would be Alzheimer's disease: there is an increasing medical need, and a few therapies are appearing on the market, but there is, as yet, no real

Advances in Drug Discovery Techniques. Edited by Alan L. Harvey © 1998 John Wiley & Sons Ltd.
ISBN 0 471 97509 5

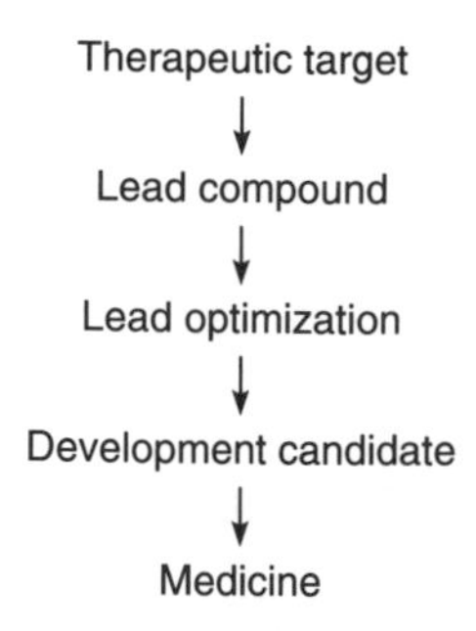

Figure 1.1 A simple flow diagram for the drug discovery and development process. The disease that is the therapeutic focus is analysed in order to provide targets that form the basis for screening. The most promising of the compounds active on the primary screens forms the lead compound, which usually goes through a phase of analogue development in order to optimize its activity and chemical properties. The optimized lead compound enters the development phase which involves toxicology testing, pharmacokinetic and metabolism studies, and clinical trials.

treatment. The second reason why drug discovery is still an important issue is because there are opportunities in many disease areas provided in the form of newly discovered therapeutic targets. These arise from fundamental research in molecular biology, and in the future new therapeutic targets are expected to result from the human genome project. The third reason for examining drug discovery processes relates to the pressures on the pharmaceutical industry. These include the world-wide trend to contain costs of medicines in various healthcare systems (James, 1994), and the need for competitive advantage, which may be gained from producing innovative products (Drews, 1995). There is also concern about the general decline of productivity, as evidenced by the increased money spent on pharmaceutical research and development but the trend towards fewer new chemical entities brought to the market each year. The latest statistics published by the Centre for Medicines Research International shows that 36 new molecular entities were introduced in 1996, compared with 61 in 1987 (Figure 1.2) (CMR, 1997). It has been estimated also that the cost of bringing a new drug to market was $500 million in 1996, compared with $231 million in 1986 (see Drews and Ryser, 1997). Overall, it is expected that premium prices will be obtained in the future only for innovative medicines that prove to meet a genuine need. One way to achieve such innovation is by discovering novel treatments and novel therapeutic targets.

1.1.2 Approaches to drug discovery

In this short introductory review, I will consider three of the current general approaches to drug discovery. The first could be called the 'sense' approach, that is rational drug design using computer aided techniques. The second could

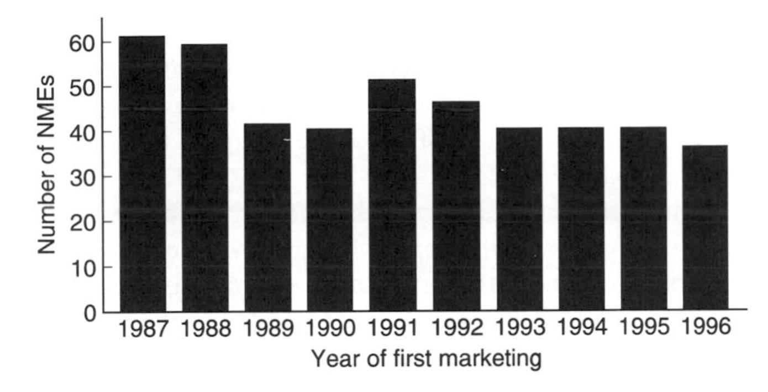

Figure 1.2 The number of new molecular entities (NMEs) introduced annually to the world market. Data from the Centre for Medicines Research International.

be called the 'nonsense' approach, that is the empirical approach of random screening, which could be regarded as 'irrational' drug design. The third major area is the 'antisense' approach, which is based on manipulation of genetic targets. Gene therapy will not be considered because it is based on manipulations that are not drug-like in their properties.

1.2 RATIONAL DRUG DISCOVERY

Rational drug discovery can be directed at postulated therapeutic targets or at known lead molecules.

1.2.1 Design based on targets

There are important limitations to this approach, mainly that the rational approach succeeds when there is a good understanding of the 3-dimensional structure of the target (see, e.g., Platt and Robson, 1992; Richards, 1992; Ringe and Petsko, 1995; Blundell, 1996). However, as many of the therapeutic targets are proteins or other large macromolecules, it is rather rare to have a good understanding of the 3D structure. Computer techniques have improved dramatically in power and sophistication in the past few years, but there are still few reliable ways to predict detailed 3D structures of a protein. This means that the structures of potential therapeutic targets have to be determined experimentally, and such techniques are still very slow. The most precise technique is x-ray crystallography, which can be a rate-limiting step in understanding protein structure. In addition, there has been a massive explosion in the number of predicted protein sequences from DNA cloning, and this will continue with future advances in molecular biology and the human genome project. To make use of this genetic information for drug design, it is necessary to be able to predict structures of proteins from sequence information.

Despite these general difficulties, there are a few examples of drug design based on therapeutic targets. Several companies have inhibitors of HIV protease, and the molecular design of some of the inhibitors was found through an understanding of the 3D structure of the target enzyme (e.g., Erickson *et al.*, 1990; Roberts and Shaw, 1993).

A second example is the small molecular weight inhibitors of thymidylate synthase which were produced by Agouron. In this case, the inhibitors were designed using predicted structure information from an enzyme in *E. coli* (Appelt *et al.*, 1991). Another example of inhibitors is the purine nucleoside phosphorylase inhibitors designed by scientists at Biocryst. Their approach was to solve the crystal structure of the PNP enzyme, both alone and in association with the putative inhibitor (Ealick *et al.*, 1991). Modified inhibitors were then synthesized, and the process was repeated until highly potent antagonists were obtained.

1.2.2 Design based on lead molecules

These can be classed according to whether the lead molecules are relatively small or large macomolecules. In the case of small molecules, there are many examples of successful drugs being derived from leads that are natural neurotransmitters or are natural products interacting at receptors or enzymes. Classical examples include the design of antagonists from naturally occuring agonists: e.g. propranolol from noradrenaline, and cimetidine from histamine (see Sneader, 1985, for a historical review; Mason, McLay and Lewis, 1995). Enzyme inhibitors with useful pharmaceutical characteristics have been derived also from starting points that include small peptides. For example, the ACE inhibitors were developed from studies on the bradykinin potentiating peptides found in the venom of the snake *Bothrops jararaca* (Cushman *et al.*, 1980).

When the lead compound is a protein or other complex macromolecule, the situation is more difficult (e.g., Hirschmann, Smith and Sprengeler, 1995; Roques, 1995). While there are empirical examples, rational drug design is limited by the difficulties highlighted earlier of predicting protein structures from sequence information (Kuntz, Meng and Shoichet, 1995; Blundell, 1996). Some attempts have been made to shortcut this bottleneck through analysis of families of related proteins in order to define small regions of the total structure that are important functionally (e.g., see Cardle and Dufton, 1994). Although still at an experimental stage, this approach led to the design of serine protease inhibitors with specificity that could be targetted between trypsin, chymotrypsin and kallikrein (Dufton, M. J. and Harvey, A. L. unpublished). The inhibitors had molecular weights of around 500, but the structural information was derived from the Kunitz family of serine protease inhibitors, whose members have molecular weights of around 7000.

1.3 'IRRATIONAL' DRUG DISCOVERY

Historically, the vast majority of clinically used drugs have been developed from empirical observations, often based on the pharmacological effects of natural products. While the experimental test systems available to detect potentially useful therapeutic activity are based on animal models of disease, the ability to screen large numbers of compounds is limited because of the time and expense involved. With recent technological developments in molecular biology, instrumentation, and information technology, it is possible to conduct screening of compounds at throughputs that could barely have been dreamt about a few years ago (Broach and Thorner, 1996). The availability of molecular targets, the ability to engineer such targets into simple 'reporter' systems (see, e.g., Hobden and Harris, 1992), and the use of robotics to handle the samples and conduct the assays make random screening of chemical diversity a very attractive approach to the discovery of novel activity. The advantages of molecular assays are summarized in Table 1.1. The challenges that have to be overcome with this approach relate to the selection of new targets and to the availability of sufficient new molecules to make high-throughput screening economically feasible.

1.3.1 New targets

The techniques of molecular cloning provide the possibility of deriving an understanding of physiological processes at the molecular level. Thirty years ago, receptors were still a theoretical concept, without a clear physical basis. Now, over 250 gene products relating to major neurotransmitters are known (see Bowman and Harvey, 1995; Watson and Girdlestone, 1996) and many different subtypes of ion channel also have been characterized genetically. There has been a similar increase in the understanding of intracellular signalling pathways, opening up the possibility of new target sites for drugs.

The molecular approach should enable a molecular dissection of any disease process. However, in practice this is unlikely to be simple: the reductionist approach loses the systems integration that is a key feature of many physio-

Table 1.1 Properties of molecular assays

Subtypes of receptor available
Human receptors available
Suitable for low-abundance proteins
Linked to simple expression sytems
Feasible for miniaturization
Suitable for automation
High throughput

logical and pathophysiological processes. The paradox is that molecular biology provides an embarrassment of riches, in that the number of potential therapeutic targets being discovered is greater than the number that can be validated experimentally. Target validation becomes a potential bottleneck for high-throughput screening. One approach to this problem is to use molecular engineering to determine whether a putative target molecule is essential for cellular function. This can be extended to *in vivo* models by creating genetically manipulated animals, but this is necessarily slower than high throughput screening; perhaps target validation will be performed by academic groups in collaboration with industry.

1.3.2 Compounds for high throughput screening

Over the last ten years, many biochemical assays have been adapted for use on 96-well microplates. This has enabled large increases in throughput to be achieved, but this in itself creates new problems (e.g., Harding *et al.*, 1997). The increase in the number of assays from, say, 10,000 per year to, potentially, 100,000 per day implies an enormous increase in the cost of consumables. This leads to moves to use smaller and smaller volumes, such that 384-well plates may soon become routine, and there are moves to reduce assay volumes from microlitres to picolitres.

In addition, there is the problem of finding sufficient chemical diversity to feed the screens. In addition to re-using their collection of compounds, many companies are obtaining new chemical diversity from combinatorial chemistry or from natural products. Both topics are discussed in later chapters, but some general comments will be included here.

Combinatorial chemistry is the general term for the approach to synthesizing compounds in parallel rather than sequentially (Baldwin, 1996; Hogan, 1996; Dolle, 1997; Nefzi, Ostresh and Houghten, 1997; Wilson and Czarnik, 1998). Various techniques have been developed (Table 1.2), and some of them are capable of generating vast numbers of different compounds. These tend to be based on peptides or oligonucleotides so that, although biological activity may be discovered on high-throughput screening, the active compound is unlikely

Table 1.2 Different approaches to combinatorial chemistry

Technique	*References*
Phage display peptides	Cwirla *et al.* (1990); Smith and Petrenko (1997)
Peptides on beads	Lam *et al.* (1991)
Peptides on chips	Fodor *et al.* (1991)
Peptides on teabags	Houghten *et al.* (1991)
Organic chemistry libraries	Nefzi, Ostresh and Houghten (1997)

to have physiochemical properties suitable for use as a drug. In the last few years, most of the developments in combinatorial chemistry have concentrated on the use of small organic building blocks, such as benzodiazepine or other heterocyclic nuclei (e.g., Gordeev *et al.*, 1996; Boojamra *et al.*, 1997; Nefzi *et al.*, 1997), in order to create libraries with more drug-like qualities (see Chapter 9 by Shuttleworth).

An additional problem relates to handling the data from the synthetic processes and determining which molecule in a mixture of related compounds is the active component (Hassan *et al.*, 1996; Leland *et al.*, 1997). This process ('deconvolution') can slow the overall discovery effort significantly, and more attention has been paid to methods for producing single compounds rather than mixtures (e.g., Lam, Lebl and Krchnak 1997; Pirrung, 1997), or on 'encoding' the location of individual molecules (Burbaum *et al.*, 1995; Pirrung, 1997).

The other major source of chemical diversity for screening purposes is natural products. These have been the basis for many clinically successful drugs (see Table 1.3), but there are more recent examples of natural products introduced into the market, e.g., the antimalarial artemisinin, and the anticancer agent taxol. Of the twenty top selling drugs in 1996, eight are of natural product origin or were derived from natural product leads. In a recent survey by Cragg, Newmann and Snader (1997), it was estimated that 39% of all 520 new approved drugs in 1983–1994 were natural products or derived from natural products, and 60–80% of antibacterial and anticancer drugs were derived from natural products. Some examples are given in Table 1.3, and more details can be found in Chapters 2 (Waterman), 3 (Hill), 5 (Stafford) and 6 (Jaspars).

1.4 THE ANTISENSE APPROACH

The aim of antisense drug therapy is to affect a pathophysiological process at the genetic level. This involves using modified oligonucleotides that bind specifically to target sequences of RNA to cause translational arrest, or to sections of DNA to block transcription (Stein and Cheng, 1994; Crooke, 1995;

Table 1.3 Drugs based on natural products

Traditional examples	*Recent examples*
Atropine	Artemisinin
Ephedrine	Atracurium
Morphine	Captopril
Physostigmine	Ivermectin
Quinine	Pravastatin
Tubocurarine	Taxol

Matteucci and Wagner, 1996; Breaker, 1997). Antisense drugs are discussed in more detail in Chapter 12 by Bennett *et al.*, but some brief comments are included here.

With the increase in our understanding of the genetics of many diseases, it is tempting to expect the future therapies will be based on genetic manipulations. Gene therapy may be applicable to diseases associated with single gene defects, but problems associated with targeting and delivery largely remain to be solved. Using the more drug-like approach of antisense oligonucleotides seems more appealing. However, there are still a number of technical problems impeding the development of antisense therapeutics. These include the size of the molecule needed to ensure specificity of action (11–15 mers seem to be the minimum size of antisense oligonucleotide); the chemical stability needs to be enhanced through the use of modified oligonucleotides; uptake of the rather large antisense molecules and distribution to the appropriate target site within the cells of the target tissue; and cost (the reagents are not cheap, and the chemistry is not trivial).

Despite these difficulties, several companies have been created in order to develop antisense therapies. A number of clinical trials are in progress (Table 1.4), although no antisense therapeutic agent has yet reached the market.

1.5 CONCLUSIONS

Drug discovery continues to be necessary in order to find more effective treatments for many important diseases. Scientific advances combine with commercial and political pressures to encourage pharmaceutical companies to improve their drug discovery processes.

The overall activity will continue to develop in a highly interdisciplinary way, and it seems likely from the technical developments discussed briefly in

Table 1.4 Antisense therapies in clinical trial[a]

Indication	*Stage*	*Compound*	*Company*
Antiviral	Phase III	Fomivirsen	ISIS
Anti-HIV	Phase II	GEM-91	Hybridon
Anticancer	Phase II	Anti-Bcl2	Genta
Crohn's disease	Phase II	ISIS-2302	ISIS
Restenosis	Phase II	LR-3280	Lynx
Anti-CMV	Phase II	GEM-132	Hybridon
Anticancer	Phase I	CGP-64128	Novartis/ISIS
Anticancer	Phase I	AML	Lynx
Anticancer	Phase I	ISIS-5132	ISIS
Anticancer	Phase I	Anti-C-myb	Lynx

[a]Information compiled from Pharmaprojects, August 1997.

this chapter and more fully in the rest of the book that modern drug discovery will no longer be a limiting factor in the development of novel medicines. Instead, the bottlenecks will be in characterizing the lead molecules produced from high throughput screening or chemical design processes. The equivalent high throughput processes need to be developed for drug absorption, distribution and metabolism, and for toxicology. Until then, successful drug discovery and development will continue to require what has always been an important factor: luck or serendipity.

REFERENCES

Appelt, K., Bacquet, R.J., Barlett, C.A, Booth, C.L.J., Freer, S.T., Fuhry, M.A.M., Gehring, M.R., Herrmann, S.M., Howland, E.F., Janson, C.A., Jones, T.R., Kan, C.C., Kathardekar, V., Lewis, K.K., Marzoni, G.P., Matthews, D.A., Mohr, C., Moomaw, E.W., Morse, C.A., Oatley, S.J., Ogden, R.C., Reddy, M.R., Reich, S.H., Schoettlin, W.S., Smith, W.W., Varney, M.C., Villafranca, J. Ernest, Ward, R.W., Webber, S., Webber, S.E., Welsh, K.M. and White, J. (1991) Design of enzyme inhibitors using iterative protein crystallographic analysis, *Journal of Medicinal Chemistry*, **34**, 1925–34.

Baldwin, J.J. (1996) Design synthesis and use of binary encoded synthetic chemical libraries, *Molecular Diversity*, **2**, 81–8.

Blundell, T.L. (1996) Structure-based drug design, *Nature*, **384**, Suppl., 23–6.

Boojamra, C.G., Burow, K.M., Thompson, L.A. and Ellman, J.A. (1997) Solid-phase synthesis of 1,4-benzodiazepine-2,5-diones. Library preparation and demonstration of synthesis generality, *Journal of Organic Chemistry*, **62**, 1240–56.

Bowman, W.C. and Harvey, A.L. (1995) The discovery of drugs, *Proceedings of The Royal College of Physicians Edinburgh*, **25**, 5–24.

Breaker, R.R. (1997) *In vitro* selection of catalytic polynucleotides, *Chemical Reviews*, **97**, 371–90.

Broach, J.R., and Thorner, J. (1996) High-throughput screening for drug discovery, *Nature*, **384**, Suppl., 14–16.

Burbaum, J.J., Ohlmeyer, M.H.J., Reader, J.C., Henderson, I., Dillard, L.W., Li, G., Randle, T.L., Sigal, N.H., Chelsky, D. and Baldwin, J.J. (1995) A paradigm for drug discovery employing encoded combinatorial libraries, *Proceedings of the National Academy of Sciences of the USA*, **92**, 6027–31.

Cardle, L. and Dufton, J.J. (1994) Identification of important functional environs in protein tertiary structures from the analysis of residue variation in 3D-application to cytochromes-*c* and carboxypeptidase-A and carboxypeptidase-B, *Protein Engineering*, **7**, 1423–31.

CMR (1997) New molecular entities: Japanese companies see a further decline in output, *Centre for Medicines Research International News*, **15**, 10–13.

Cragg, G.M., Newmann, D.J. and Snader, K.M. (1997) Natural products in drug discovery and development, *Journal of Natural Products*, **60**, 52–60.

Craig, A.-M. and Malek, M. (1995) Market structure and conduct in the pharmaceutical industry, *Pharmaceutics and Therapeutics*, **66**, 301–37.

Crooke, S.T. (1995) Oligonucleotide therapeutics, in *Burger's Medicinal Chemistry and Drug Discovery*, 5th Edn, Vol. 1, *Principles and Practice* (ed. M.E. Wolff), Wiley, New York, pp. 863–900.

Cushman, D.W., Ondetti, M.A., Cheung, H.S., Sabo, E.F., Antonaccio, M.J. and Rubin, B. (1980) Angiotensin-converting enzyme inhibitors, in *Enzyme Inhibitors and Drugs* (ed. M. Sandler), Macmillan, London, pp. 231–47.

Cwirla, S.E., Peters, E.A., Barrett, R.W. and Dower, W.J. (1990) Peptides on phage: a vast library of peptides for identifying ligands, *Proceedings of the National Academy of Sciences of the USA*, **87**, 6387–92.

Dolle, R.E. (1997) Discovery of enzyme inhibitors through combinatorial chemistry, *Molecular Diversity*, **2**, 223–36.

Drews, J. (1995) *The Impact of Cost Containment on Pharmaceutical Research and Development*, Centre for Medicines Research, Carshalton.

Drews, J. and Ryser, S. (1997) Pharmaceutical innovation between scientific opportunities and economic constraints, *Drug Discovery Today*, **2**, 365–72.

Ealick, S.E., Babu, Y.S., Bugg, C.E., Erion, M.D., Guida, W.C., Montgomery, J.A. and Secrist, J.A. (1991) Application of crystallographic and modeling methods in the design of purine nucleoside phosphorylase inhibitors, *Proceedings of the National Academy of Sciences of the USA*, **88**, 11540–4.

Erickson, J., Neidhart, D.J., Vandrie, J., Kempf, D.J., Wang, X.C., Norbeck, D.W., Plattner, J.J., Rittenhouse, J.W., Turon, M., Wideburg, N., Konlbrenner, W.E., Simmer, R., Helfrich, R., Paul, D.A. and Knigge, M. (1990) Design, activity, and 2.8 Å crystal structure of a C_2 symmetric inhibitor complexed to HIV-1 protease, *Science*, **249**, 527–33.

Fodor, S.P.A., Read, J.L., Pirrung, MC., Stryer, L., Lu, A.T. and Solas, D. (1991) Light-directed, spatially addressable parallel chemical synthesis, *Science*, **251**, 767–73.

Gordeev, M.F., Patel, D.V., Wu, J., Gordon, E.M. (1996) Approaches to combinatorial synthesis of heterocycles: solid phase synthesis of pyridines and pyrido[2,3-d]pyrimidines, *Tetrahedron Letters*, **37**, 4643–6.

Harding, D., Banks, M., Fogarty, S. and Binnie, A. (1997) Development of an automated high-throughput screening system: a case history, *Drug Discovery Today*, **2**, 385–90.

Hassan, M., Bielawski, J.P., Hempel, J.C., Waldman, M. (1996) Optimization and visualization of molecular diversity of combinatorial libraries, *Molecular Diversity*, **2**, 64–74.

Hirschmann, R., Smith III, A.B. and Sprengeler, P.A. (1995) Some interactions of macromolecules with low molecular weight ligands. Recent advances in peptidomimetic research, in *New Perspectives in Drug Design* (eds P.M. Dean, G. Jollies and C.G. Newton), Academic Press, London, pp. 1–12.

Hobden, A.N. and Harris, T.J.R. (1992) The impact of biotechnology and molecular biology on the pharmaceutical industry, *Proceedings of the Royal Society of Edinburgh*, **99B**, 37–45.

Hogan, J.C. (1996) Directed combinatorial chemistry, *Nature*, **384**, Suppl., 17–19.

Houghten, R.A., Pinilla, C., Blondelle, S.E., Appel, J.R., Dooley, C.T. and Cuervo, J.H. (1991) Generation and use of synthetic peptide combinatorial libraries for basic research and drug discovery, *Nature*, **354**, 84–6.

James, B.G. (1994) *The Pharmaceutical Industry in 2000. Reinventing the Pharmaceutical Company*. The Economist Intelligence Unit, London.

Kuntz, I.D., Meng, E.C. and Shoichet, B.K. (1995) Challenges in structure-based drug design, in *New Perspectives in Drug Design* (eds P.M. Dean, G. Jollies and C.G. Newton), Academic Press, London, pp. 137–50.

Lam, K.S., Lebl, M. and Krchnak, V. (1997) The "one-bead-one-compound" combinatorial library method, *Chemical Reviews*, **97**, 411–48.

Lam, K.S., Salmon, S.E., Hersh, E.M., Hruby, V.J., Kakzmierski, W.M. and Knapp, R.J. (1991) A new type of synthetic peptide library for identifying ligand-binding activity, *Nature*, **354**, 82–4.

Leland, B.A., Christie, B.D., Nourse, J.G., Grier, D.L., Carhart, R.E. (1997) Managing the combinatorial explosion, *Journal of Chemical Information and Computer Sciences*, **37**, 62–70.

Mason, J.S., McLay, I.M. and Lewis, R.A. (1995) Application of computer-aided drug design techniques to lead generation, in *New Perspectives in Drug Design* (eds P.M. Dean, G. Jollies and C.G. Newton), Academic Press, London, pp. 225–50.

Matteucci, M.D. and Wagner, R.W. (1996) In pursuit of antisense, *Nature*, **384**, Suppl., 20–2.

Nefzi, A., Ostresh, J.M. and Houghten, R.A. (1997) The current status of heterocyclic combinatorial libraries, *Chemical Reviews*, **97**, 449–72.

Nefzi, A., Ostresh, J.M., Meyer, J.P. and Houghten, R.A. (1997) Solid phase synthesis of heterocyclic compounds from linear peptides: cyclic ureas and thioureas, *Tetrahedron Letters*, **38**, 931–4.

Pirrung, M.C. (1997) Spatially addressable combinatorial libraries, *Chemical Reviews*, **97**, 473–88.

Platt, E. and Robson, R. (1992) Case studies in automatic modelling of thrombin, alpha-lactalbumin and other proteins, and implications for drug design, *Proceedings of the Royal Society of Edinburgh*, **99B**, 123–36.

Richards, W.G. (1992) Computer-aided drug discovery, *Proceedings of the Royal Society of Edinburgh*, **99B**, 105–11.

Ringe, D. and Petsko, G.A. (1995) The age of structure: the role of protein crystallography in drug design, in *New Perspectives in Drug Design* (eds P.M. Dean, G. Jolles and C.G. Newton), Academic Press, London, pp. 89–112.

Roberts, N.A. and Shaw, S. (1993) Discovery and development of the HIV proteinase inhibitor Ro31-8959, in *The Search for Antiviral Drugs* (eds J. Adams and V.J. Merluzzi), Birkhauser, Boston.

Roques, B.P. (1995) Drug design based on structural similarity and molecular biology, in *New Perspectives in Drug Design* (eds P.M. Dean, G. Jollies and C.G. Newton), Academic Press, London, pp. 15–33.

Smith, G.P. and Petrenko, V.A. (1997) Phage display, *Chemical Reviews*, **97**, 391–410.

Sneader, W. (1985) *Drug Discovery: the Evolution of Modern Medicines*, Wiley, Chichester.

Stein, C.A. and Cheng,Y.-C. (1994) Antisense oligonucleotides as therapeutic agents – is the bullet really magical?, *Science*, **261**, 1004–12.

Watson, S. and Girdlestone, D. (1996) Trends in Pharmacological Sciences, Suppl., *Receptor and Ion Channel Nomenclature*, Elsevier, Cambridge.

Wilson, S.R. and Czarnik, A.W. (1998) *Combinatorial Chemistry: Synthesis and Application*, Wiley, Chichester.

2

Natural Products for Drug Discovery: an Overview

Peter G. Waterman

The cells of living organisms – plants, fungi, bacteria, lichens, insects, and higher animals – are the sites of intricate and complex synthetic activities that result in the formation of a remarkable array of organic compounds, many of them of great practical importance to mankind.

(Geissman and Crout, 1969)

2.1 INTRODUCTION

This chapter is the first of several dealing with various aspects of natural products as the origin of chemical diversity for drug discovery screening programmes: see Chapters 3 (Hill), 5 (Stafford), 6 (Jaspars) and 7 (Hunter). Therefore, this is an overview of the advantages and disadvantages of using natural products in drug discovery, and many specific aspects are dealt with in the later chapters. Most of the examples in this chapter will be from higher plants, but the general comments apply equally well to other taxonomic groups.

We are fortunate to be surrounded by organisms that possess an extraordinary capacity to manipulate relatively simple building blocks ('precursors') into an enormous structural array of small 'drug-like' molecules. Often these are referred to as secondary metabolites because they possess no obvious primary metabolic function. Just how many different species of organism there are is still a matter for debate, but figures of as high as 30 million are being quoted and, in many cases and notably among microorganisms, a species may well be made up of many strains with differing metabolic profiles. We must remember also that the organisms in existence today represent the culmination

Advances in Drug Discovery Techniques. Edited by Alan L. Harvey © 1998 John Wiley & Sons Ltd.
ISBN 0 471 97509 5

of perhaps 300 million years of chemical evolution in which natural selection pressures seem to have been directed toward maximizing both structural diversity and biological activity (Harborne, 1988).

As observed by Geissman and Crout (1969), the cells of plants are the sites for intricate and complex biosynthetic activity that results in the formation of a remarkable array of metabolites, a high proportion of which are characterized by possessing one to several chiral centres. This chirality imparts a 3-dimensional component into many metabolites and makes them difficult for the synthetic chemist to copy. Yet one of the most fascinating features of the chemistry of nature is that secondary metabolites are generated, in all their diversity, from a very small number of precursors. Most notable among the precursors are acetyl coenzyme A, mevalonic acid and the aromatic amino acids and their non-nitrogenous derivatives arising from the shikimic acid pathway. These three, together with less commonly used intermediates such as the non-aromatic amino acids and simple hexose and pentose sugars, are deployed over and over again in many different arrangements and combinations to construct this vast array of chemical complexity. Today the number of known structures stands at over 120,000 (Buckingham, 1997).

2.2 SECONDARY METABOLITES

2.2.1 Their *raison d'être*

Why do living organisms invest so much resource into the production of secondary metabolites? This has long been a matter for debate, and while the issue is still contentious it is clear that the original idea that they were waste products can now be largely discounted.

Without being able to neatly define or wholly explain function it is now widely believed that the role of the majority of secondary metabolites is to contribute to the producer's 'fitness'. That is, through interaction with the environment or with other organisms in the environment, secondary metabolites enhance the prospects for survival of the producer or its offspring. Among the various potential roles they may play in fulfilling this definition is in the defence of the producer against predators or pathogens. To cite but one example, there is now an extensive literature implicating secondary metabolites produced by higher plants in the defence of their producers against herbivory (Rosenthal and Berenbaum, 1991). While higher plants, because of their sedentary lifestyle, tend only to be able to exploit metabolites in a defensive capacity, this is far from true of all organisms. Metabolites can equally well be offensive weapons: for example, pathogen to plant, or animal against animal. Venoms from snakes and spiders and the conotoxins from sea snails are a very interesting and complex group of offensive substances that exhibit a diverse array of pharmacological activity.

Thus, many of the compounds classified as secondary metabolites have evolved to exhibit biological activity against at least some other organism or group of organisms in the environment. From the drug discovery perspective, this is a useful prerequisite.

2.2.2 Structural diversity of secondary metabolites within a species

Given that what is critical to the random screening process is diversity in chemical structures, an important question becomes what level of structural diversity occurs in an organism? Unfortunately any answer to this question is bound to be hypothetical because the exhaustive isolation of all compounds in an extract is a practical impossibility. Where commercial importance has led to an extract being the focus of intense analytical activity (for example, alcoholic and caffeine-containing beverages, volatile oils, and a few pharmaceuticals, notably opium), it is clear that the numbers of isolatable compounds are in the hundreds. As separation and identification techniques improve, that number will continue to grow. Furthermore, experience suggests that at the species level, and often within a species (variety, strain of microorganism), the chemistry of an organism will be unique.

The arguments put forward above will not always be supported in research. Natural product chemists studying species as an academic exercise generally focus on a particular type of compound in which they specialize, or employ isolation techniques that are optimized to their expectations of what will be present. These factors will lead to an increased chance of finding what is anticipated and a consequent reduction in the chance of finding the 'unexpected'.

Where the detector guiding the isolation process is a bioassay rather than a particular extraction protocol this can change, and even well studied species will provide unusual and novel compounds. Take, for example, the isolation of the antifertility compound yuehchukene (**1**) from *Murraya paniculata* (Rutaceae) by means of a bioassay guided procedure (Kong *et al.*, 1985). Prior to the discovery of **1**, this species had been reported, on numerous occasions, as a source of 8-prenylated-7-oxygenated coumarins and polyoxygenated flavonoids (more than 30 publications) but alkaloids like yuehchukene were previously unknown.

Another facet of secondary metabolites that is useful to the drug discovery process is their propensity to occur in 'families' of allied compounds. Consequently, a bioassay-guided separation commonly can yield a series of structurally allied bioactives. These effectively represent a small combinatorial library, the study of which is likely to throw light on structure–activity relationships. A good example of such a family of compounds is the modified clerodane diterpenes found in several closely allied Flacourtiaceae from Costa Rica (Figure 2.1).

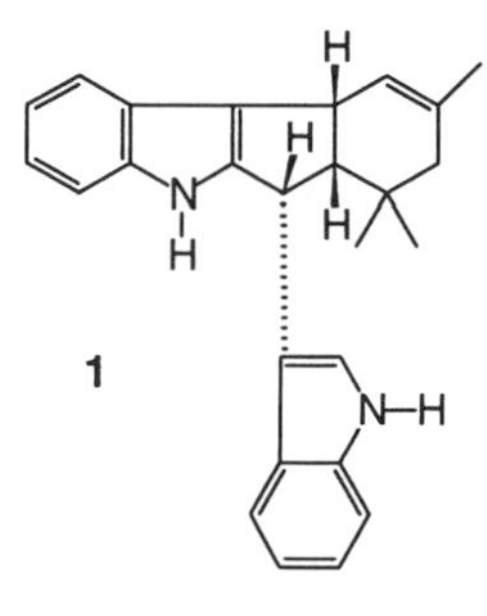

Structure 2.1

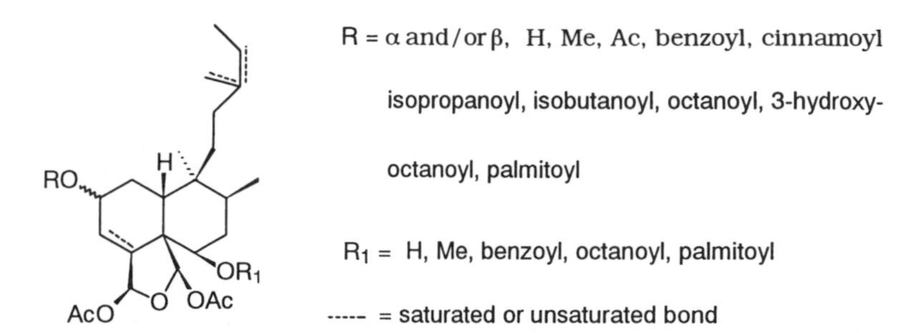

Figure 2.1 Clerodane diterpenes from Costa Rican Flacourtiaceae (Khan *et al*, 1990a,b; Khan, Gray and Waterman, 1990; Gibbons, Gray and Waterman, 1996a,b).

2.3 A FUTURE FOR NATURAL PRODUCTS IN DRUG DISCOVERY?

Historically, natural products have been the major source of medicines and much of the world's population still relies upon herbal remedies. As Western pharmaceutical philosophy moved towards dealing with defined chemical entities, then isolation of active principles from herbal products gave some important drugs, such as morphine, codeine, digoxin, cocaine and quinine. Indeed, quinine is arguably one of the most important drugs ever discovered since its activity against the malarial parasite effectively allowed the European races to colonize parts of the world where they would otherwise have succumbed to malaria. Screening programmes have identified other important natural products such as the alkaloids of *Catharanthus roseus* (vinblastine, vincristine) and the vast array of antibiotics derived from *Penicillium* and *Streptomyces* spp. According to the analysts James Capel and Co., eight of the world's top 25 drugs in 1994 (in terms of worldwide sales) were derived from microbes (Chicarelli-Robinson, 1997). Yet, today we find ourselves going through a period of doubt regarding the value of natural products in drug research similar to that experienced in the late 1970's. This is despite the fact that the 1990's have seen the arrival of such important new metabolites as taxol (**2**) and artemisinin (**3**), both of which have given rise to commercial products, for challenging therapeutic areas.

The current 'problem' for natural products is the arrival of combinatorial chemistry as a cleaner and more controllable source of large numbers of structures at a time when ultra high throughput screening has come into vogue. There has been the inevitable fanfare heralding this new technology as the answer to the problem of generating sufficient numbers of compounds to satisfy the new roboticized screens of some major companies. The view has arisen that the two sources of chemical diversity are in direct competition, the consequence of which has been the downgrading or elimination of natural product programmes in some companies.

In the long term the argument of direct competition cannot actually be sustained. Nature continues to offer unrivalled and virtually random structural diversity, whereas combinatorial chemistry offers unparalleled ability to explore structure–activity relationships around a lead compound. Far from being competitive with each other, the two should be viewed as mutually supportive components in the drug discovery armoury.

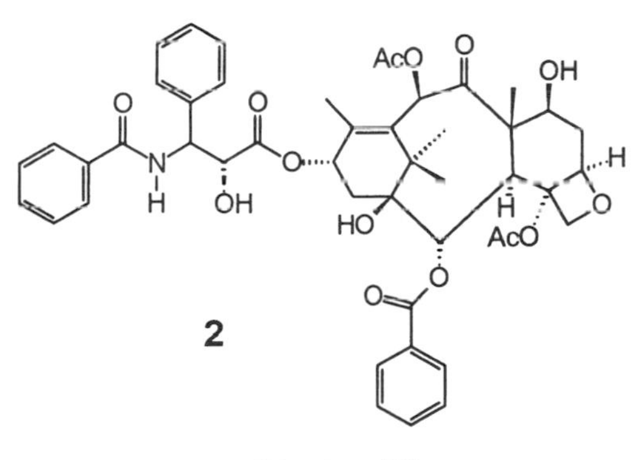

Structure 2.2

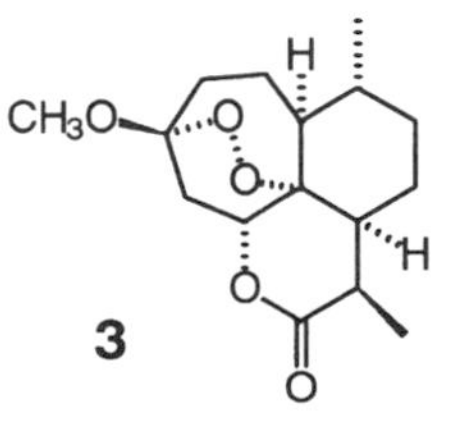

Structure 2.3

There is an important point in the above for the advocate of natural products to accept. The lead compound in a drug discovery operation may well be derived from a natural source (Buss and Waigh, 1995), but there should be relatively little expectation that this lead will find its way to market. While nature will continue to be good at providing us with the ideas for new drugs, in the vast majority of cases the efforts of medicinal chemists, increasingly through combinatorial chemistry, will identify synthetic or semisynthetic compounds that are more appropriate for use. An example from my own University has been the development of the neuromuscular blocking drug atracurium (**4**), which was conceived through an understanding of the mode of action of the alkaloid tubocurarine (Waigh, 1988).

One final point is that we should expect an increasing diversity of natural product sources to be involved in providing chemical diversity. Traditionally, fungi and plants have been the major providers, while, more recently, marine organisms, snake and spider venoms, and insects have come under investigation. The manipulation of natural biosynthetic pathways will also be explored in more detail. The strategy of stressing plant cell cultures into aberrant biosynthetic processes is discussed in Chapter 5 by Stafford. There is currently much interest in gene manipulation as a means by which to expand natural product libraries from microorganisms, particularly by adding biosynthetic genes from organisms adapted to extreme environments (see Chapter 7 by Hunter). Combinatorial strategies based on natural products also are being pursued vigorously by companies such as ChromaXome (Sherman, 1997).

2.3.1 Biodiversity and bioprospecting issues

Chemical diversity among natural products is correlated directly with biological diversity. To obtain large numbers of chemicals from natural resources it is necessary to access high biodiversity. With a few notable exceptions like

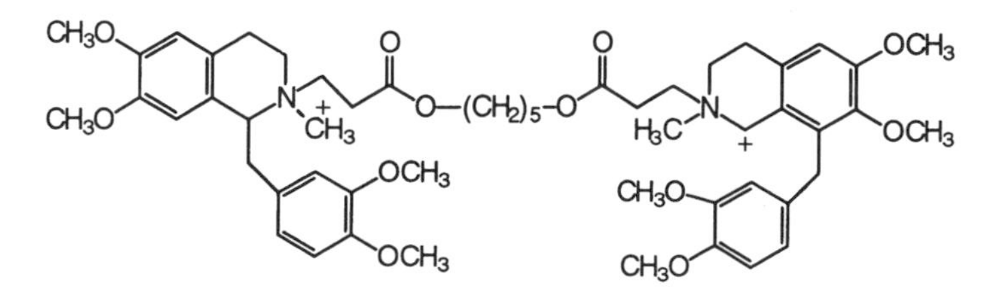

4

Structure 2.4

Australia and the states of Hawaii and Florida in the USA, areas of high biodiversity are located in the developing world. Access to biodiversity for use in drug discovery programmes (bioprospecting) has become a major issue in the last few years (Reid *et al.*, 1993). The source countries have become conscious of many years of uncontrolled collecting of materials without any mechanism for recognizing the exploitation of resources or dealing with recompense (short and long term) for that exploitation. There has been a serious polarization of views between the owners of the biodiversity who, quite understandably, look upon it as a valuable resource and the potential resource users who see relatively little value in the raw material until an exploitable bioactivity has been identified. A few pharmaceutical companies have taken positive action to combat this problem, most notably Merck through its agreement with the Instituto Nacional de Bioversidad (INBio) in Costa Rica. By contrast, others have either tried to continue to access material without commitment or have viewed the difficulty and potential adverse publicity involved in such interactions as another argument for not using natural products.

The United Nations Convention on Biological Diversity attempts to put the issues into an agreed framework, but relatively few countries have implemented the Convention. Some source countries are now developing legislation to control the use of their biodiversity. This is a two-edged sword. If handled pragmatically, and with an understanding of the realities of the drug discovery process, a clear and workable set of ground rules will encourage the participation of pharmaceutical companies, but if they are too complex and have unrealistic expectations for up-front commitments it will discourage. Examples of both situations are now becoming apparent.

In the Strathclyde Institute for Drug Research we have become involved, through our international network of contacts, in brokering the supply of biodiversity for screening under licence. By linking ourselves with our suppliers through collaborative agreements that are financially transparent we have been able to build up a library of extracts for licensing and for use in our own drug discovery programme. Some of the more developed among the SIDR collaborations are planned to go further in the future, with the establishment of jointly run drug discovery research centres with appropriate technology transfer.

2.4 TECHNICAL PROBLEMS IN USING NATURAL PRODUCTS IN RANDOM SCREENING

At its most simple the screen detects bioactivity in the extract, the active compound(s) are then separated and purified using the biological screen as a detector, and finally the purified active compound(s) are identified. In each of these three phases there are currently important developments underway which promise to improve greatly the efficiency of these processes. Some of these are highlighted below.

2.4.1 Sample preparation

Crude extracts are very complex mixtures that contain both desirable and undesirable compounds. Generally they are prepared by solvent extraction (hot or cold) from either solid biomass or a fermentation broth. The type of compound extracted can be controlled to some degree by the extraction protocol.

There is no right or wrong protocol, and the most appropriate method is likely to vary with the type and format of the bioassay. Typical pre-assay treatments of extracts involve removal of tannins and chlorophylls, the former because of their ability to bind with proteins and the latter because their colour often can have a deleterious effect on the measurement of activity. Simple, semi-automated, column chromatographic procedures, employing polyvinylpyrrolidone (PVP) to complex with polyphenols and a gel filtration procedure (Sephadex) to void chlorophylls can eliminate both problems.

High throughput screening usually is reliant on 96-well or 384-well plate technology and the extract team is expected now to produce their products for use in this environment. Often this can be done by producing deep-well 'mother' plates, which can then be used to generate a large number of lower concentration 'daughter' plates for screening.

A consequence of the move to high throughput technologies is that the limiting component of the assay procedure is now, more often than not, the availability of enough chemical diversity to keep the screen operating at optimal rates. This has tended to change thinking about how best to screen complex mixtures. In the past, where the assay has been rate-limiting and the goal is to screen as widely as possible, then there are obvious advantages in employing complex mixtures of compounds for each assay point. However, if the assay is not rate-limiting then there are advantages in making the diversity examined at each assay point less complex, as this will simplify subsequent activities. This has caused some companies to look at methods for fractionating complex mixtures or even complete isolation of individual compounds from mixtures prior to screening. Xenova, for example, are planning to build libraries of isolated but unidentified compounds (Chicarelli-Robinson, 1997; see Chapter 3 by Hill). A few years ago one would have viewed this type of approach as prohibitively labour intensive but today, with the advent of a whole range of automated and semi-automated fractionation methodologies based on HPLC and small column flash chromatography, the situation has been transformed (Muller-Kuhrt, 1997).

2.4.2 Fractionation of bioactive mixtures and dereplication

With the rapidly increasing number of screening events that can be anticipated from the move to ever-higher throughput screening, thc bottleneck in the drug discovery process is now the processing of hits. This can be overcome partly by employing secondary assays that may be more difficult or time consuming but which impart more information and so allow for the elimination of hits that do not entirely fulfil previously defined criteria. However, at some point, if the discovery process is going to go forward, the active compound or compounds will need to be separated.

Traditionally this was a laborious process usually based on the separation of mixtures by their elution from solid media due to their variable partition or adsorption, or by gel filtration or, more rarely, by liquid–liquid partition. Considerable improvements in most chromatographic procedures, usually involving a degree of automation, has reduced the time taken to achieve separation to the degree where one good technician or graduate can, today, accomplish as much as 10 workers would have done only 10 years ago! Claims are being made that the next generation of automated chromatographic separation will be able to separate each component of a complex mixture in a single run. The 'Sepbox' technology being developed by Analyticon (Muller-Kuhrt, 1997), which employs HPLC technology, is a most interesting example of what is now becoming possible (Figure 2.2).

A critical part of the duties of the team responsible for analysing active fractions is what has become known as the dereplication process. Dereplication is the recognition and elimination from the discovery pathway of compounds that have no potential for exploitation, perhaps because of prior knowledge of their activity or known toxicity. Dereplication processes tend to exploit LC/MS and LC/PDA systems which will, minimally, provide data for a library based on UV, MW and retention times. There is now talk about being able to add on line ^{1}H NMR analysis which, when it is done, will increase enormously the power of the dereplication process.

The earlier dereplication takes place in the pathway the greater its value. The recent move towards pre-separation and identification of compounds prior to screening could be accompanied by dereplication at that point. But this should be done with caution, because discarding a compound before it has been put to what may well be a unique screen is a risky business.

2.4.3 Identification

Originally the rate-limiting step in natural products studies was the elucidation of the actual structure, particularly where it involved any degree of chirality. Here, more than anywhere else in the various processes involved in natural products-based drug discovery, things have changed. The extraordinary rise in the power of NMR over the past 15 years means that relatively small amounts

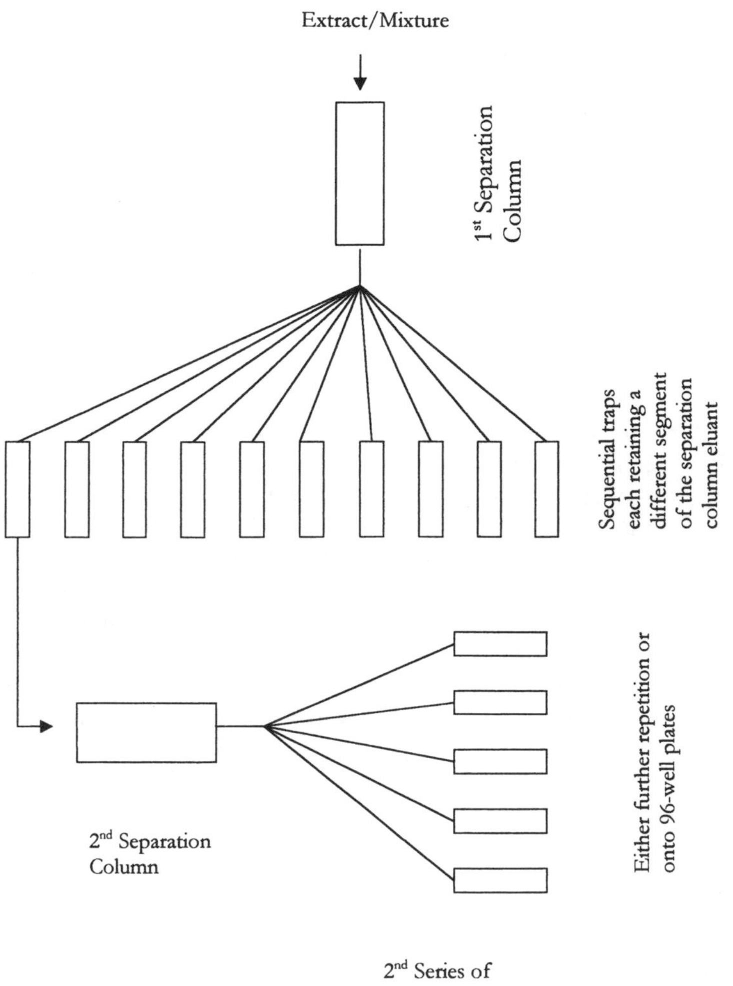

Figure 2.2 A strategy for automated fractionation of a crude extract based on HPLC. The strategy outlined is similar to the SEPBOX procedures of Analyticon (Muller-Kuhrt, 1997).

of pure compounds (1 mg) can be exposed to the full range of 1D, 2D and 3D homo- and heteronuclear procedures within a 48 hour experimental time frame. Collectively these techniques invariably will give significant insights into the structure, and more often than not will resolve it to the level of relative stereochemistry. In a less spectacular manner mass spectrometry also has become a much more versatile technique, offering a wide range of methods for observing molecular ions and fragmentation patters. Given access to state-of-the-art hardware it is now rare for these procedures not to resolve structural problems within an acceptable period of time.

REFERENCES

Buckingham, J. (1997) *Dictionary of Natural Products*, Version 3.2 on CD-Rom, Chapman & Hall, London.

Buss, A.D. and Waigh, R.D. (1995) Natural products as leads for new pharmaceuticals, in *Burger's Medicinal Chemistry and Drug Discovery* (ed. M.E. Wolff), Vol. 1, Wiley, New York, pp. 983–1033.

Chicarelli-Robinson, I. (1997) Highly competitive approach to fast track natural products drug discovery, in International Business Communications Symposium on Natural Products Drug Discovery, San Diego, 17–18 March.

Geissman, T.A. and Crout, D.H.G. (1969) *Organic Chemistry of Secondary Plant Metabolism*, Freeman Cooper, New York.

Gibbons, S., Gray, A.I. and Waterman, P.G. (1996a) Clerodane diterpenes from the bark of *Casearia tremulosa*, *Phytochemistry*, **41**, 565–70.

Gibbons, S., Gray, A.I. and Waterman, P.G. (1996b) Clerodane diterpenes from the leaves of *Laetia procera*, *Phytochemistry*, **43**, 635–8.

Harborne, J.B. (1988) *Introduction to Ecological Biochemistry*, 3rd Edn, Academic Press, London.

Khan, M.R., Gray, A.I., Reed, D.R., Sadler, I.H. and Waterman, P.G. (1990a) Diterpenes from *Zuelania guidonia*, *Phytochemistry*, **29**, 1609–14.

Khan, M.R., Gray, A.I., Sadler, I.H. and Waterman, P.G. (1990b) Diterpenes from *Casaeria corymbosa*, *Phytochemistry*, **29**, 3591–5.

Khan, M.R., Gray, A.I. and Waterman, P.G. (1990) Diterpenes from *Zuelania guidonia*, Stem bark, *Phytochemistry* , **29**, 2939–42.

Kong, Y.-C., Cheng, K.-F., Cambie, R.C. and Waterman, P.G. (1985) Yuehchukene: A novel indole alkaloid with anti-implantation activity, *Chemical Communications*, 47–8.

Muller-Kuhrt, L. (1997) Automated isolation as the key approach to make natural product screening more competitive, in International Business Communications Symposium on Natural Products Drug Discovery, San Diego, 17–18 March.

Reid, W.V., Laird, S.A., Meyer, C.A., Gamez, R., Sittenfeld, A., Janzen, D.H., Gollin, M.A. and Juma, C. (1993) *Biodiversity Prospecting: Using Genetic Resources for Sustainable Development*, World Resources Institute, USA.

Rosenthal, G.A. and Berenbaum, M.R. (1991) *Herbivores: Their Interactions with Secondary Plant Metabolites*, 2nd Edn, Academic Press, New York.

Sherman, D. (1997) Combinatorial biology and the natural product continuum, in International Business Communications, Post Conference Workshop on Interfacing Combinatorial and Natural Products Technologies, San Diego, 18 March.

Waigh, R.D. (1988) The chemistry behind atracurium, *Chemistry in Britain*, 1209–12.

3

Advanced Screening Technology and Informatics for Natural Products Drug Discovery

Diana C. Hill

3.1 INTRODUCTION

The importance of microbial natural products as sources of new drugs began to be recognized in the 1940s and 1950s when the impact of the discovery of penicillin became apparent (Berdy, 1974, 1980). Since this discovery, natural resources have produced important enzyme inhibitors such as clavulanic acid (Reading and Cole, 1977), a β-lactamase inhibitor which is used in combination with β-lactam antibiotics to overcome resistance in antibacterial therapy, and mevinolin, a hydroxymethylglutaryl-CoA reductase inhibitor for treatment of hypercholesterolaemia (Albert *et al.*, 1980). Other pharmacologically significant inhibitors include cyclosporin A (Borel *et al.*, 1976) and FK506 (Kino *et al.*, 1987), which are important immunosuppressive agents.

For the discovery of new drugs from microbial natural products a diverse and novel culture collection is required (Yarborough *et al.*, 1993). At Xenova we have capitalized on the ability of microorganisms and plants to produce chemical diversity by building a culture collection of 26 000 microbes and up to 7000 plant species, focusing on those from rare and unusual habitats.

For the microbes, the biosynthetic potential of the organisms is exploited using fermentation conditions that are targeted to stimulate production of secondary metabolites. This production is enhanced by using different nitrogen and carbon sources and other diverse medium components, and by the use of solid substrates for the growth of higher fungi such as the basidomycetes.

Advances in Drug Discovery Techniques. Edited by Alan L. Harvey © 1998 John Wiley & Sons Ltd.
ISBN 0 471 97509 5

For the plants, materials are obtained from diverse geographical habitats and metabolites are extracted from different plant parts such as the roots, leaves and shoots separately.

To ensure that Xenova's chemical (NatChem^{TM}) library produced from the microbial and plant collections is optimal for screening, stringent extraction procedures are carried out. These are designed to maximize the yield of secondary metabolites while minimizing the presence of common interferences such as high molecular weight material that will result in a false positive result in the screen.

Once the library has been prepared it is then screened to detect active samples or 'hits' (see section 3.2.2) which are analysed in detail prior to selection of high quality samples for further work. This analysis is performed by a proprietary informatics system which is described in section 3.3.2.

Following hit selection, purification and structure elucidation of active metabolites is carried out using a range of chromatographic and spectroscopic techniques. During this process, biological profiling of active fractions and compounds is used to ensure the potency and specificity of the new chemical templates isolated (Figure 3.1).

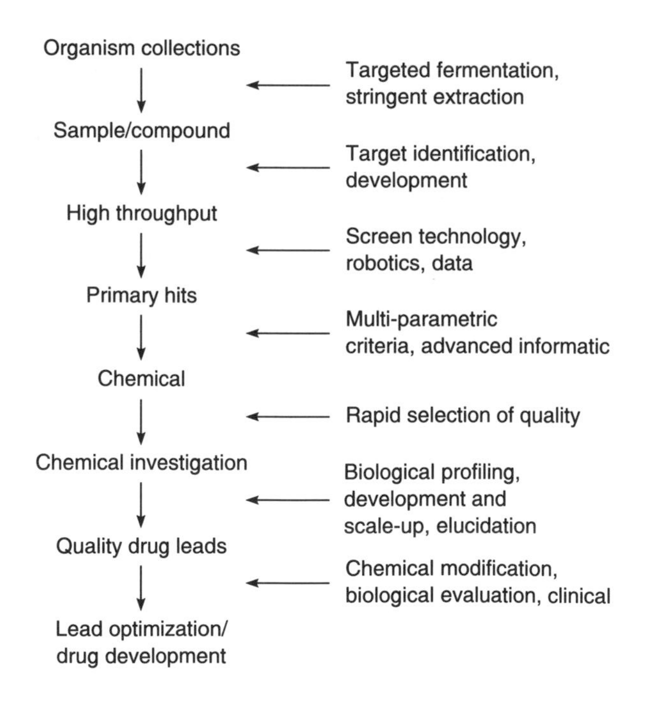

Figure 3.1 Overview of procedures involved in natural products drug discovery.

3.2 SCREENING TECHNOLOGY

3.2.1 Requirements of screens for natural products drug discovery

Screens designed to test natural products must be sensitive, selective and able to test large numbers of samples (Yarborough *et al.*, 1993). Use of appropriate technology to achieve a lower detection limit of 10–200 nM is important. This is because the concentration of metabolites in each library sample is unknown, so it is important to be able to detect potent compounds that are present in low concentrations. Assuming that active metabolites are present at concentrations of 1–10 μg/ml in the crude extract have an average molecular weight of 500 Da and are diluted 20–200-fold in the assay, the required detection limit of the screen is in the 10–200 nM range.

Screens should be specific also for the molecular or cellular disease target of choice. Appropriate discriminatory tests, such as cytotoxicity measurement for cell-based assays, or isotype specificity tests for molecular assays, are important as they add value to the data obtained for screening hits. In addition, the data generated from all screens in which samples have been tested should be compared so that selective hits can be identified at an early stage. This combination of specific screens, data comparisons and discriminatory assays makes possible the selection of the best hits for further work.

The screening system must work in the presence of the library to be tested, and therefore must be compatible with its physicochemical characteristics. Hence, natural products screens are operational in the presence of solvents, are buffered against extremes of pH and ionic strength, and are not affected by colour.

Finally, screens should be carried out in the presence of suitable controls to measure their performance and usually should be semi- or fully automated to enable rapid screening of libraries. The advantages and disadvantages of using modular automation for high throughput screening programmes have been reviewed by Hook (1995).

3.2.2 Review of screening technology

A broad range of technologies have been used for screening programmes (Figure 3.2). For molecular targets these range from solution phase assays (Gopalakrishna *et al.*, 1992; Walker, Winder and Kellam, 1993) to immobilized substrate assays (Farley *et al.*, 1992, Sadick *et al.*, 1995) and generic techniques such as scintillation proximity assays (SPA) and time resolved fluorescence (TRF) which are described below. For cell-based targets, cell-signalling (Brann *et al.*, 1996) and reporter gene assays (Dhundale and Goddard, 1996) have been used widely.

Generally, screening technology is moving towards the use of assays with a minimum number of steps. One example is a novel assay for the detection of

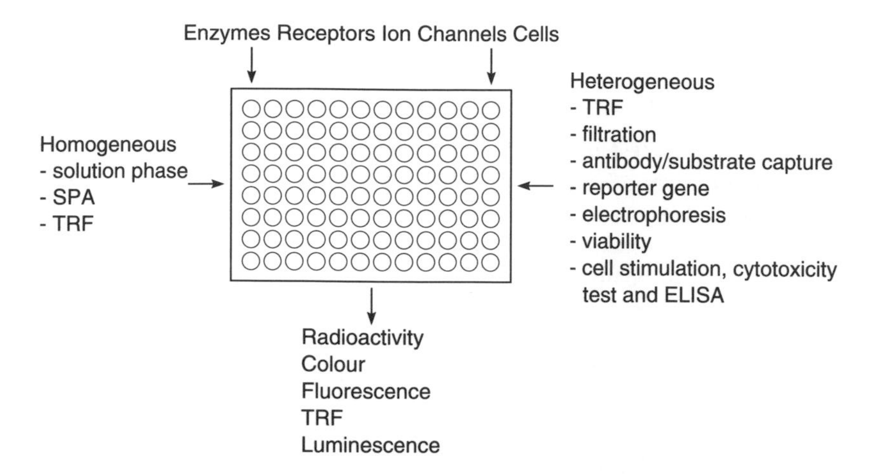

Figure 3.2 Commonly used screening technologies. Inputs include assay reagents generated from different sources and design of an appropriate assay format for use in testing large numbers of samples (SPA, scintillation proximity assay; TRF, time resolved fluorescence; ELISA, enzyme linked immunosorbent assay). Outputs are determined using microtitre plate readers using the technologies shown, and compared with controls to determine whether any inhibition of the signal is observed in the presence of test samples.

inhibitors of the TNFα receptor-ligand interaction (MacAllan *et al.*, 1997). In this assay, cloned TNFα receptor is used to coat the surface of microtitre plates. Test samples and ^{125}TNFα are added to the plates, followed by incubation for 1 hour, then the plates are washed and counted using a microtitre plate liquid scintillation counter. This assay has been developed into a rapid and robust screen that is amenable to automation.

SPA (Udenfriend *et al.*, 1985) is a homogeneous radioisotopic technique which relies on the limited path length a β particle will travel through aqueous media. If the electron collides with a scintillant particle, energy is transferred and light is emitted. In SPA, scintillant is incorporated into beads or coated on to, or incorporated in, the microtitre plate itself. This makes it possible to carry out a range of assays (Taylor *et al.*, 1994, Lerner *et al.*, 1996) including cell-based assays for monitoring *in situ* receptor binding, cellular metabolism, and cell motility (Cook, 1995). SPA assays are amenable to automation, and indeed have been completely automated, as described by McCaffrey, Powers and Kelly Talbot (1996).

TRF utilizes the properties of lanthanides (europium (Eu), terbium (Tb), samarium (Sm) and others) to overcome problems such as background fluorescence and quenching that often are seen with other fluorometric techniques. One application of TRF is dissociation enhanced lanthanide fluorescence immunoassay (DELFIA) which is appropriate for use in heterogeneous assays (i.e., those requiring a separation step).

In DELFIA, Eu is attached to one of the biological reagents (antibody, protein, nucleic acid, hapten) to be used in the screens via a chelating compound (Hemmila and Harju, 1995). When bound to the biological reagent, the fluorescence of the Eu is quenched by the surrounding water molecules, and practically no background fluorescence is observed. Following screen completion, an enhancement solution at low pH which contains β-NTA (a fluorogenic chelating ligand), TOPO (to replace water molecules) and Triton X100 (to solubilize the assay components and provide optimized conditions for fluorescence) is added to the screen and the Eu dissociates from the biological reagent. The Eu is captured in a micelle that excludes water and the lanthanide begins to fluoresce with a high intensity. Eu fluorescence produces a sharp emission peak, has a long decay time and exhibits a large Stokes shift (i.e., the difference between the excitation and the emission wavelength), as shown in Figure 3.3. This means that fluorescence can be measured after background has decayed and that assays are

• Dissociation enhancement results in high fluorescence intensity

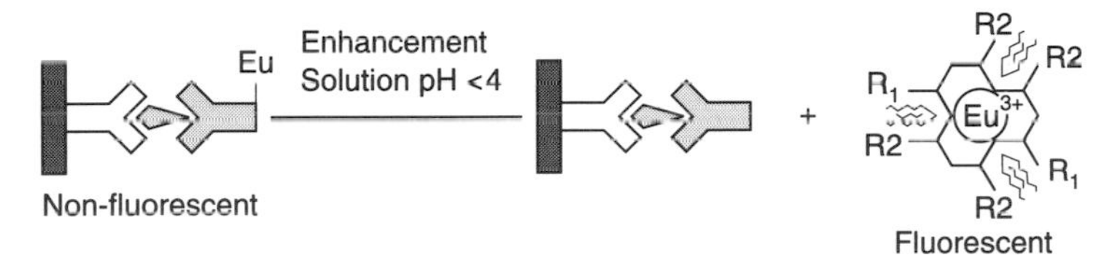

• Time resolved measurement, large Stokes' shift and sharp emission peak eliminates background

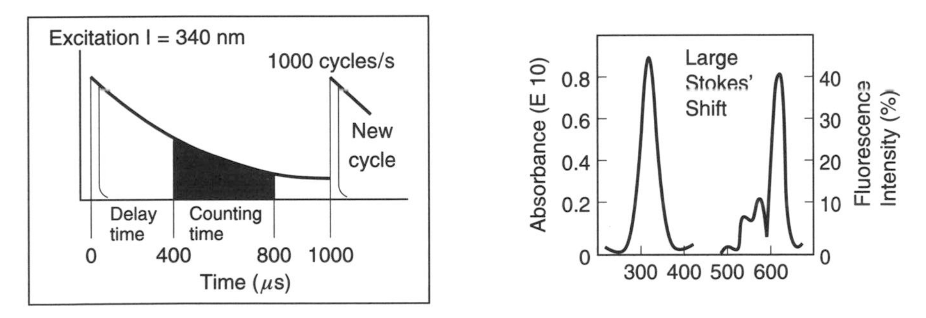

Figure 3.3 Principles of lanthanide TRF. Biological reagents are labelled with a lanthanide that is non-fluorescent in the presence of water. Following assay completion the lanthanide is dissociated from the reagent and fluorescence enhanced as described in the text. The properties of the lanthanides eliminate background fluorescence which commonly has been a problem with other fluorescence techniques.

highly sensitive and have a wide dynamic range (Dickson, Pollak and Diamandis, 1995).

DELFIA assays can be applied across a range of assay formats including enzyme, antigen–antibody, protein–protein and DNA–protein assays (Figure 3.4). The stable signal generated, together with the low fluorescence background, make the technology ideal for automation. Importantly, multiple label assays using more than one lanthanide for measurement of two analytes, or inclusion of an internal control to enhance the quality of data produced (Hurskainen, McAllan and Hill, 1996), are possible.

Another application of TRF is its use in homogeneous or non-separation assays due to its properties in delayed fluorescence resonance energy transfer (DEFRET, Figure 3.5). In this application, one biological reagent is labelled with a lanthanide donor molecule and the other with an energy acceptor molecule. When the two reagents are in proximity (approximately 30 Å), energy is transferred from the lanthanide to the acceptor molecule and fluorescence is emitted at a different wavelength. If the binding of the two molecules is inhibited, no energy transfer occurs. Different high throughput screen formats using homogeneous TRF have been published (Kolb, Yamanaka and Manly, 1996; MacAllan, Hurskainen and Hill, 1996).

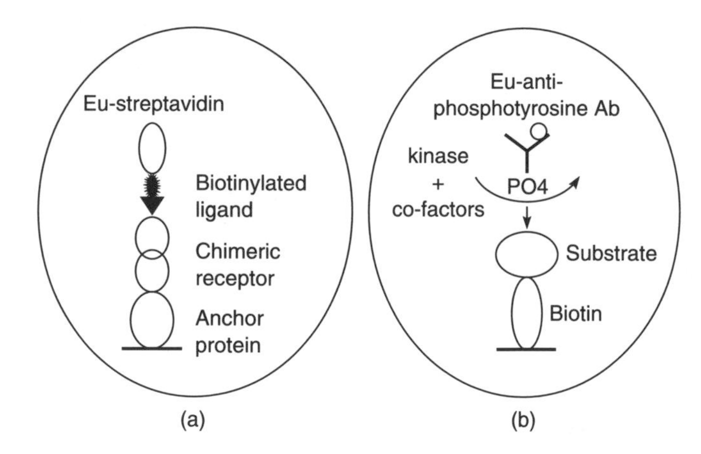

Figure 3.4 Applications of DELFIA (dissociation enhanced lanthanide fluorescence immunoassay). (a) Receptor–ligand binding assay where the receptor is bound to the surface of a microtitre plate using an anchor protein and a biotinylated ligand added with the sample. If no inhibition of binding occurs a signal can be generated from the assay using europium labelled streptavidin. Inhibition of the receptor–ligand interaction results in a reduced fluorescence reading. (b) Protein tyrosine kinase assay where a biotinylated substrate is attached to a streptavidin coated microtitre plate. The enzyme is added with the sample and phosphorylation of the substrate on tyrosine residues can be measured using europium labelled anti-phosphotyrosine antibody.

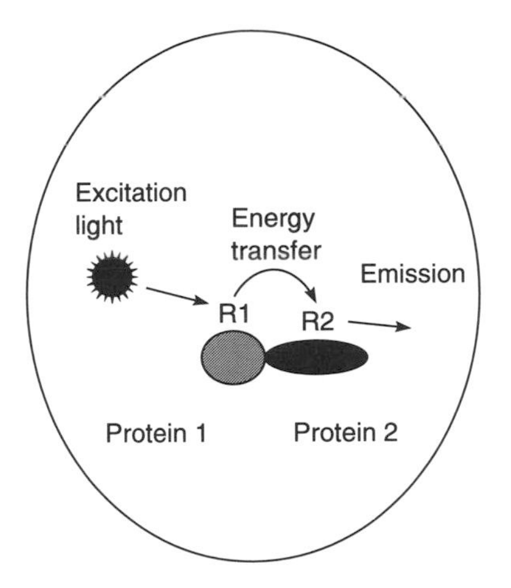

Figure 3.5 Homogeneous time resolved fluorescence assay. The assay reagents are labelled with compounds capable of accepting and transferring energy (RI, R2). When the two reagents are within 30 Å of one another and an appropriate excitation wavelength is used, energy transfer occurs between R1 and R2 which can be measured at the emission wavelength. If the interaction between the reagents is inhibited by test samples the energy emitted is reduced.

A further homogeneous fluorescence technique is fluorescence polarization. In this case, the ligand is labelled with a fluorescent tag and used in a technique, based on the rotation of molecules, which measures the decrease in fluorescence depolarization that is exhibited by small molecules when their rotation is slowed by binding to large molecules (Checovich, Bolger and Burke, 1995). The quantification of biomolecular interactions then can be carried out using competition methods, making this technique useful as a high throughput screening method (Sportsman *et al.*, 1995, Jolley, 1996).

A fluorescence technique designed specifically to perform quantitative optical screening for cell-based kinetic assays has recently become available. The fluorescent imaging plate reader (FLIPR) which reads plates in this technique incorporates low-level optical detection and precise temperature control and liquid handling, to allow measurement of kinetics in a 96-well microtitre plate in under 1 second. Demonstrated applications of this technique include measurements of intracellular calcium, intracellular pH and membrane potential (Schroeder and Neagle, 1996).

These examples demonstrate clearly the move towards use of fluorescence as a detection technology in the pharmaceutical industry, and the increasing preference for homogeneous assays which are faster, enable kinetic studies to be carried out and facilitate the move towards miniaturization. Currently miniaturization is being exploited by a number of small biotechnology companies, such as Aurora Biosciences and Evotec.

3.3 ADVANCED INFORMATICS FOR NATURAL PRODUCTS DRUG DISCOVERY

3.3.1 Database requirements and design

For natural products drug discovery to be successful, the disciplines involved must be closely integrated (Yarborough *et al.*, 1993) both operationally and in terms of data exchange. The important role of informatics systems in areas of drug discovery such as target identification, characterization of drug activity (Dhar *et al.*, 1996) and analysis of quantitative structure–activity relationships (Maddalena, 1996) is recognized.

The principles of such systems have been combined at Xenova to produce a powerful bio- and chemo-informatics system (RAPIDD™) to store and analyse the large quantities of diverse data produced from the key natural products discovery areas of library generation and large scale metabolite production, screening, and purification and structure elucidation of active metabolites. When considering the design of the informatics system our overall objective was to make maximum use of the data generated in all areas with equal weighting. At the time (late 1993) there were no systems available commercially that were capable of storing and handling the data generated from the different disciplines involved. Therefore we worked with programming consultants (Sechi Information Systems Ltd) to design a system that can store and analyse data from internal and external sources, link and search large amounts of data rapidly, and enable us to reduce time to lead discovery by focusing our resources on productive targets.

The resulting informatics system covers the areas previously described and includes also an area of the database which defines which discipline the user is from and, very importantly, management and reporting functions (Figure 3.6). The key benefits of the database storage and organization functions are that operator entry of data is minimal, electronic exchange of data is available to all, and transfer of information by paper is reduced significantly.

3.3.2 Screening informatics

Two key areas are involved in the screening process. The first is development and validation of the assay, and the second is high throughput screening and directing the purification of active metabolites (Figure 3.7).

In assay development, suitable assay reagents and technology are evaluated and an assay format is established. The screen database has free format areas to allow the data generated from these investigations to be stored. Following development, rigorous validation procedures are applied to the assay, assessing its reproducibility, compatibility with the physicochemical characteristics of the samples and converting it to a screen. This last mentioned process involves investigation of operational parameters such as incubation times, wash steps

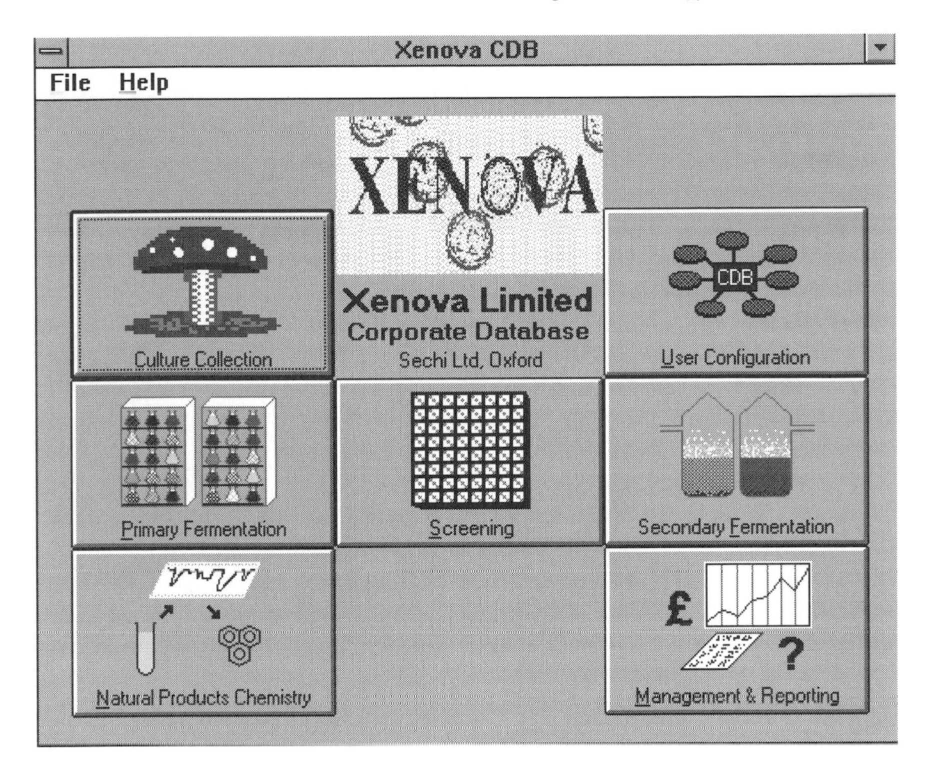

Figure 3.6 Xenova's proprietary informatics system.

and instrumentation to be used. It also includes pilot screening of approximately 700 compounds and 1000 organisms that broadly represent the chemical and biological diversity of the NatChem™ library. Pilot screening makes possible the estimation of hit rates and the determination of data analysis procedures to be carried out for the full screening programme. Valuable information on the chemical classes that may produce activity is obtained also from this study, information that later can be used for the prioritization of samples for chemistry.

During high throughput screening and direction of the purification of active metabolites, data are generated from plate readers or the robotic system, retests of active samples (known as 'Xens' for purposes of analysis using the database) are carried out with dilution, and active fractions or compounds are tested in the screen. Cross screen data (i.e., from all the screens in which the sample has been tested) and discriminatory assay data are used prior to selection of samples for chemistry. The database therefore provides the facility for the automatic generation of flexible plate plans for testing of the different samples and inclusion of appropriate controls. The resulting data can be analysed plate by

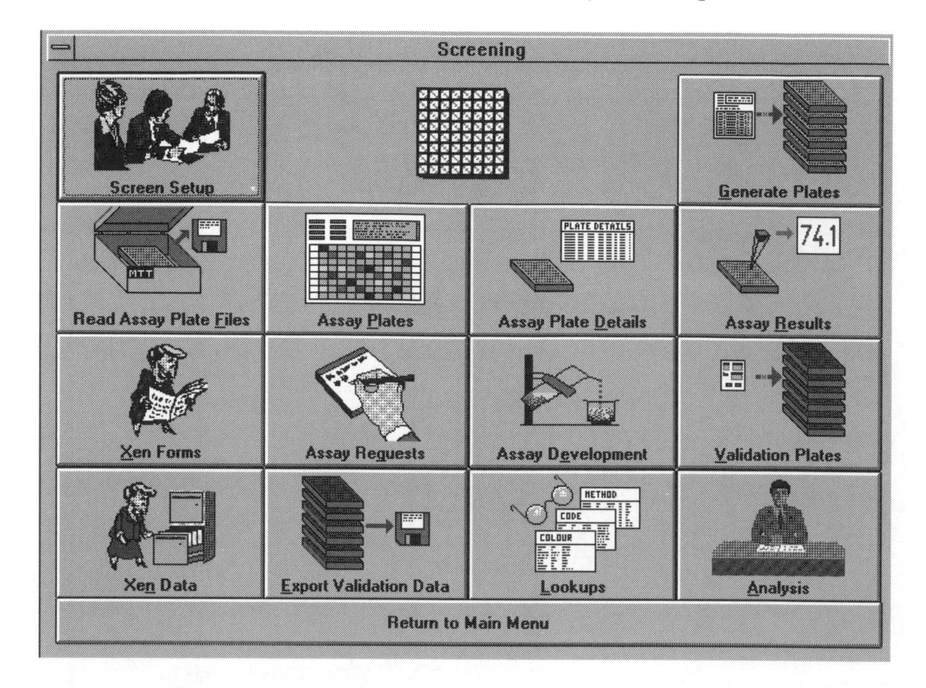

Figure 3.7 Screening informatics. The front screen shows functions related to high throughput screening (Screen Setup: SOPs, assay, operator, assay details; Generate Plates: electronic generation of plate formats for sample testing; Read Assay Plate Files: loading of data into database from plate readers; Assay Plates/Details/Results: functions to allow data to be visualized, Xen Forms/Data: functions to allow multidisciplinary data to be drawn together for active samples; Assay Requests: ability for other groups to request assays: Assay Development, Validation Plates, Export Validation Data: functions to allow storage and analysis of data from development and validation activities; Lookups/Analysis: data analysis functions.

plate or across data series. This is important for natural product drug discovery because different organism types or media may produce differences in the baseline activity observed when analysing the distribution profiles of activity in each screen. In addition, by selecting an individual microtitre plate well, information on organism source, taxonomy and fermentation, together with metabolites previously isolated from extracts of the organism, can all be retrieved quickly. This multidisciplinary information is valuable also when making decisions about which samples to select for further work.

3.3.3 Integrated applications

Using RAPIDD™ it is possible to carry out detailed query searching of the information in all areas of the database. Customized management information systems are available also, and these can be used together with a project planning facility to track the progress of each project in depth. These important functions provide for detailed work scheduling and focusing of resources as required, thereby enhancing the productivity of discovery of new chemical templates.

3.4 NEW DRUG TEMPLATES FROM NATURAL PRODUCTS

To September 1996, the combination of the technologies described at Xenova has led to the discovery of 326 compounds with novel biological activity, using over 50 screens and screening more than 7 million samples. Of these, 123 compounds were novel and represented 55 different chemical classes.

Examples of novel compounds that have been published include an acidic chromone-substituted xanthone active against CD4 binding (Gammon *et al.*, 1994), novel azophilones that inhibit endothelin binding (Pairet *et al.*, 1995) and the GABA-benzodiazepine inhibitory xenovulenes (Ainsworth *et al.*, 1995). In addition, a series of inhibitors of the tissue plasminogen activator (tPA)/plasminogen activator inhibitor-1 (PAI-1) interaction is in advanced preclinical programmes, and inhibitors of multi-drug resistance are in Phase 1 clinical trials. Drug lead evaluation and development also is ongoing for several partnership programmes.

To further enhance the productivity of new drug discovery from natural products an additional approach currently is being pioneered together with corporate partners. In this approach, a library of approximately 10 000 pure natural products (QTC™) is being prepared by building on the data in the chemo-informatics system established for NatChem™ screening. This approach will bring the average cycle time of a natural products drug discovery project down to equal those of chemical and combinatorial library screening, while maintaining the yield of broad and unique chemical diversity that has been demonstrated using the NatChem™ system.

REFERENCES

Albert, S.A.W., Chen, J., Kuron, G., Hunt, V., Huff, J., Hoffman, C., Rothtock, J, Lopez, M., Joshua, H., Harris, E., Patchett, A., Monoghan, R., Currie, S., Stapley, E., Albers-Schonberg, G., Hensens, O., Hurshfield, J., Hoogsteen, K., Liesch, J. and Springer, J. (1980) Mevinolin: a highly potent competitive inhibitor of hydroxymethylglutaryl-coenzyme A reductase and a cholesterol-converting agent, *Proceedings of the National Academy of Sciences of the USA*, **77**, 3957–61.

Ainsworth, A.M., Chicarelli-Robinson, M.I., Copp, B.R., Fauth, U., Hylands, P.J., Holloway, J.A., Latif, M., O'Beirne, G., Porter, N., Renno, D.V., Richards, M. and Robinson, N. (1995) Xenovulene A, a novel GABA-benzodiazepine receptor binding compound produced by *Acremonium strictum*, *Journal of Antibiotics*, **48**, 568–73.

Berdy, J. (1974) Recent developments of antibiotic research and classification of antibiotics according to chemical structure, *Advances in Applied Microbiology*, **18**, 309–406.

Berdy, J. (1980) Recent advances in and prospects of antibiotic research, *Process Biochemistry*, **15**, 28–35.

Borel, J.F., Fevrer, C., Gubler, H.U. and Stahelin, H. (1976) Biological effects of cyclosporin A, a new antilymphocytic agent, *Agents and Actions*, **6**, 468–75.

Brann, M.R., Messier, T., Dorman, C. and Lannigan, D. (1996) Cell-based assays for G-protein coupled/tyrosine kinase-coupled receptors, *Journal of Biomolecular Screening*, **1**, 43–5.

Checovich, W.J., Bolger, R.E. and Burke, T. (1995) Fluorescence polarisation – a new tool for cell and molecular biology, *Nature*, **375**, 254–6.

Cook, N.D. (1995) Latest advances using scintillation proximity assay, in *Rapid Screening Technologies for the Discovery of Novel Drug Candidates*, Conference Proceedings, IBC Technical Services, London.

Dhar, S., Nygren, P., Soka, K., Botling, J., Nilson, K. and Larsson, R. (1996) Anticancer drug characterisation using a human cell line panel representing defined types of drug resistance, *British Journal of Cancer*, **74**, 888–96.

Dhundale, A. and Goddard, C. (1996) Reporter assays in the high throughput screening laboratory: a rapid and robust first look?, *Journal of Biomolecular Screening*, **1**, 115–18.

Dickson, E.F.G., Pollak, A. and Diamandis, E.P. (1995) Ultrasensitive bioanalytical assays using time resolved fluorescence detection, *Pharmacology and Therapeutics*, **66**, 207–35.

Farley, K., Mett, H., McGlynn, E., Murray, B. and Lydon, N.B. (1992) Development of solid phase enzyme-linked immunosorbent assays for the determination of epidermal growth factor receptor and $pp60^{c\text{-}src}$ tyrosine protein kinase activity, *Analytical Biochemistry*, **203**, 151–7.

Gammon, G., Chandler, G., Depledge, P., Elcock, C., Wrigley, S., Moor, J., Cammanota, G., Sinigaglia, F. and Moore, M. (1994) A fungal metabolite which inhibits the interaction of CD4 with major histocompatibility complex-encoded class II molecules, *European Journal of Immunology*, **24**, 991–8.

Gopalakrishna, R., Chen, Z.H., Gundimeda, U., Wilson, J.C. and Anderson, W.B. (1992) Rapid filtration assays for protein kinase C activity and phorbol ester binding using multiwell plates with fitted filtration discs, *Analytical Biochemistry*, **206**, 24–35.

Hemmila, I. and Harju, R. (1995) Time-resolved fluorometry, in *Bio-analytical Applications of Labelling Technologies* (eds I. Hemmila, T. Stahlberg and P. Mottram), Wallac Oy, Turku, pp. 83–120.

Hook, D.J. (1995) The advantages and disadvantages of using modular automation in chemical library sample distribution and the development of assays for high throughput screening programs, in Proceedings of the 2nd European Conference on High Throughput Screening and on Exploiting Molecular Diversity, Society for Biomolecular Screening, Banbury, USA.

Hurskainen, P., MacAllen, D. and Hill, D.C. (1996) Development of a novel homogeneous time resolved fluorescence energy transfer assay for the TNFα-receptor interaction, in Proceedings of the 2nd Annual Conference of the Society for Biomolecular Screening, Society for Biomolecular Screening, Danbury, USA.

Jolley, M.E. (1996) Fluorescence polarisation assays for the detection of proteases and their inhibitors, *Journal of Biomolecular Screening*, **1**, 33–8.

Kino, T., Hatanaka, H., Hashimoto, M., Nishiyama, M., Goto, T., Okuhara, M., Koshaka, M., Aoki, H. and Imanaka, H. (1987) FK-506, a novel immunosuppressant isolated from a *Streptomyces* I. Fermentation, isolation and physico-chemical and biological characteristics, *Journal of Antibiotics*, **40**, 1249.

Kolb, J.M., Yamanaka, G. and Manly, S.P. (1996) Use of a novel homogeneous fluorescent technology in high throughput screening, *Journal of Biomolecular Screening*, **1**, 203–10.

Lerner, C.G., Chiang Saiki, A.Y., Mackinnon, A.C. and Xuei, X. (1996) High throughput screen for inhibitors of bacterial DNA topoisomerase I using the scintillation proximity assay, *Journal of Biomolecular Screening*, **1**, 135–43.

MacAllan, D., Hurskainen, P. and Hill, D.C. (1996) Development of a novel dual lanthanide labels TNFα-receptor binding assay for screening NatChem™ libraries, in Proceedings of the 2nd Annual Conference of the Society for Biomolecular Screening, Society for Biomolecular Screening, Danbury, USA.

MacAllan, D., Sohal, J., Walker, C., Hill, D.C. and Moore, M. (1997) Development of a novel TNFα-ligand-receptor binding assay for screening NatChem™ libraries, *Journal of Receptor and Signal Transduction Research*, **17**, 521–9.

Maddalena, D.J. (1996) Applications of artificial neural networks to quantitative structure–activity relationships, *Expert Opinion in Therapeutic Patents*, **6**, 239–51.

McCaffrey, C., Powers, G. and Kelly Talbot, M. (1996) Utilization of the ORCA to completely automate scintillation proximity assays, *Journal of Biomolecular Screening*, **1**, 187–90.

Pairet, L., Wrigley, S.K. Chetland, I., Reynolds, E.E., Hayes, M.A., Holloway, J., Ainsworth, A.M., Katzer, W., Chen, X.M., Hupe, D.J., Charlton, P. and Doherty, A.M. (1995) Azaphilones with endothelin receptor binding activity produced by *Penicillium sclerotiorum*: taxonomy, fermentation, isolation, structure elucidation and biological activity, *Journal of Antibiotics*, **48**, 913–23.

Reading, C. and Cole, M. (1977) Clavulanic acid: a beta-lactamase inhibiting beta-lactam from *Streptomyces clavuligeris*, *Antimicrobial Agents and Chemotherapy*, **11**, 852–7.

Sadick, M., Beresini, M., Yen, R., Golloway, A., Yeh, S. and Wong, W.L. (1995) High throughput screening assays for small molecule receptor agonists/antagonists, in Proceedings of the 2nd European Conference on High Throughput Screening and on Exploiting Molecular Diversity, Society for Biomolecular Screening, Danbury, USA.

Schroeder, K.S. and Neagle, B.D. (1996) FLIPR: a new instrument for accurate, high throughput optical screening, *Journal of Biomolecular Screening*, **1**, 75–80.

Sportsman, R., Bukar, R., Lee, S. and Dilley, H. (1995) Fluorescence polarisation assays for high throughput screening, in Proceedings of the 2nd European Conference on High Throughput Screening and on Exploiting Molecular Diversity, Society for Biomolecular Screening, Danbury, USA.

Taylor, P.B., Culp, J.S., Debouck, C., Johnson, R.K. Patil, A.D., Woolf, D.J., Brooks, I. and Herzberg, R.P. (1994) Kinetic and mutational analysis of human immunodeficiency virus type I reverse transcriptase inhibition by inophyllums, a novel class of non-nucloside inhibitors, *Journal of Biological Chemistry*, **269**, 6325–31.

Udenfriend, S., Gerber, L.D., Brink, L. and Spector, S. (1985) Scintillation proximity radioimmunoassay utilising ^{125}I-labelled ligands, *Proceedings of the National Academy of Sciences of the USA*, **82**, 8672–6.

Walker, J.M., Winder, J.S. and Kellam, S.J. (1993) High throughput microtitre plate-based chromogenic assays for glycosidase inhibitors, *Applied Biochemistry and Biotechnology*, **38**, 141–6.

Yarborough, G.G., Taylor, D.P., Rowlands, R.T., Crawford, M.S. and Lasure, L.L. (1993) Screening microbial metabolites for new drugs – theoretical and practical issues, *Journal of Antibiotics*, **46**, 535–44.

4

Discovery of Specific Tyrosine Kinase Inhibitors from Microbial and Botanical Sources

Gerald McMahon, Flora M. Tang, Tamotsu Furumai, K. Peter Hirth and Toshikozu Oki

4.1 DRUG DISCOVERY AND EXTRACTS FROM NATURAL SOURCES

Natural products or derivatives of natural products are the basis for most antimicrobial, immunosuppressive, cholesterol-lowering, and anticancer drugs on the market today to treat human diseases. This success is due, in part, to the rich chemical diversity that is found in extracts from microbial (actinomycetous, bacterial and fungal) and botanical sources. Traditionally, the identification of chemical leads from natural product sources frequently was a long and laborious process where active components in crude extracts were identified following procedures to fractionate extracts into individual chemical components. Bioassays that identified active compounds often were complex processes utilizing bacteria, fungi, or mammalian cells grown in solid-support or liquid cultures. Individual compounds then were prepared in quantities in order to demonstrate efficacy in animal models of human disease. For screening, pharmacological endpoints using cellular effects (cytotoxicity, growth inhibition) were used. Mainly it was compounds that represented the major chemical species in a given extract that were tested most often. While this approach has been and continues to be an active approach in drug discovery, the predominant chemicals in a given extract may represent only a subset of the chemical diversity that is present. The identification of minor or trace chemical entities, however, may be complicated by the requirement of the active

Advances in Drug Discovery Techniques. Edited by Alan L. Harvey © 1998 John Wiley & Sons Ltd.
ISBN 0 471 97509 5

component to have sufficient specific activity to score in the screen bioassay compared with other non-specific or unwanted activities. Using complex bioassays, it is not surprising that mechanistic studies to elucidate the drug mode of action often have lagged behind the utility of many drugs.

The use of target-driven drug discovery has made possible the rapid identification of chemicals from natural product sources that were under-represented or masked by more predominant chemicals. In this case, a molecular target serves as a basis to identify chemical entities with the desired features to be tested later in disease-relevant assays involving cells and animals. In some cases, this process allows also for the testing of extracts using assays that are diagnostic for relative specificity (Figure 4.1). In the case of chemical mixtures derived from natural sources, this process allows one to focus interest on those extracts that may have the desired specificity requirements and favour the likelihood of more useful chemical lead compounds for preclinical evaluation. In the past, identification of useful antimicrobial (antibacterial and antifungal) extracts was amenable to evaluation for relative specificity by a comparison of effects in the target cell with effects on related (different genera) and non-related (mammalian) cells. For target-driven drug discovery, the use of specificity or cross screen assays has a clear rational advantage. However, it was not known whether such an approach would be useful for our objective and provide a sufficient number of potentially useful extracts for further evaluation. The use of large numbers of highly diverse extracts from fungal and plant sources would be advantageous and favour success. In addition, our sources of microbial and plant extracts were of sufficient quality to perform the studies. The use of high-throughput screens and related counter-screens allowed us to process extracts efficiently to assess the viability of this approach. Using extracts derived from microbial and botanical sources, we sought to identify specific inhibitors of tyrosine kinases.

4.2 INHIBITORS OF TYROSINE KINASES FROM NATURAL SOURCES

Tyrosine kinases are implicated as critical molecular targets that regulate a wide variety of cellular processes including growth, differentiation, metabolism, and survival. It is not suprising, therefore, that specific and abnormal tyrosine kinase function has been implicated in a large spectrum of human diseases including cancer (growth, survival, neo-vascularization, metastases), psoriasis, fibrotic diseases (pulmonary, kidney and liver), acute transplant rejection, atherosclerosis, chronic vascular disease, rheumatoid and osteoarthritis, and arterial restenosis (Plowman, Ullrich and Shawver, 1994) In the area of cancer therapeutics, a substantial amount of effort has been directed at the identification of anticancer agents from microbial, plant, and marine sources (Chang and Geahlen, 1992). Anticancer agents that possess inhibitory properties of tyrosine and serine/threonine kinases have provided new chemical

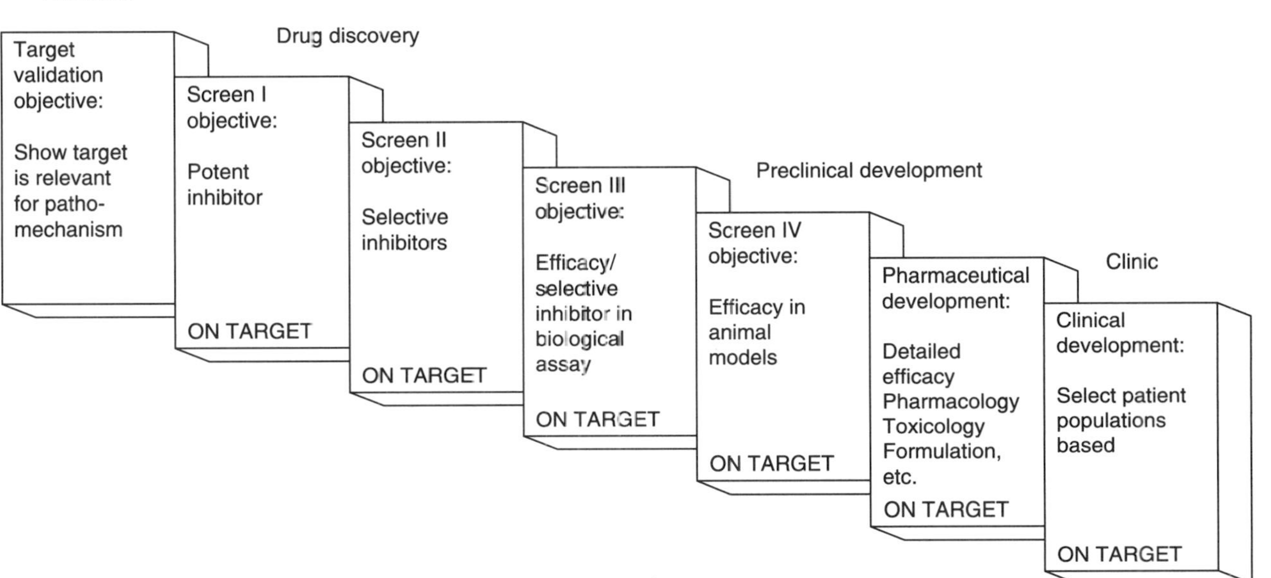

Figure 4.1 Target-driven drug discovery.

entities for the therapeutic evaluation in cancer patients. In this regard, flavopiridol, an inhibitor of cdk-2 (cyclin dependent kinase-2) currently is under clinical evaluation by the National Cancer Institute and the National Institutes of Health, Bethesda, Maryland, USA (Christian *et al.*, 1997). In the past, kinase inhibitors were derived from microbial sources and represented a wide variety of distinct chemical classes (Figure 4.2). However, many of these compounds have not been successful as drug candidates due to the lack of specific inhibitory effects on particular kinase types or due to other pharmacologic constraints that limit their utility in animal models and human clinical trials. Nonetheless, such chemicals have served as a basis for the generation of an expanding number of synthetic compounds that show potent and, in some cases, specific inhibitory properties of particular kinases (Fry *et al.*, 1994; Levitzki and Gazit, 1995; Groundwater *et al.*, 1996). Moreover, tyrosine and serine/threonine kinase inhibitors continue to be identified from natural sources (Powis *et al.*, 1994; Slate *et al.*, 1994; Rasouly and Lazarovici, 1994; Zhang *et al.*, 1995; Alvi *et al.*, 1997).

4.3 IMPLEMENTATION OF HIGH THROUGHPUT CELL-BASED RTK ASSAYS

The purpose of this study was to determine whether we could implement a method to identify specific inhibitors of receptor tyrosine kinases (RTKs) using dried extracts provided in a 96-well microtitre format from microbial (Toyama Prefectural University, Biotechnology Research Center, Toyama, Japan) and botanical (Phytopharmaceutical Laboratory, Institute of Botany, Academia Sinica, 20 Xanxin Cun, Beijing, 10093 P.R. China) sources. In pilot studies, we have determined that cell-based RTK assays are an effective means for identifying specific and potent inhibitors with synthetic chemical collections using single discrete compounds. In this study, we implemented three RTK assays that we felt would be effective for discriminating specific inhibitors that affect the intracellular catalytic core of the receptor. As shown in Figure 4.3, assays corresponding to the EGF (epidermal growth factor), PDGF (platelet derived growth factor), and IGF-1 (insulin-like growth factor-1) receptor tyrosine kinases were utilized since they represented distinct and unrelated subfamilies of tyrosine kinases. In addition, we implemented cell based assays to ensure that the active inhibitory components were membrane permeable, since the catalytic core of the enzyme is intracellular. This feature of the assay would also tend to eliminate highly toxic extracts and exclude membrane impermeable macromolecules.

The assay design used mouse fibroblasts that had been engineered to express elevated levels of the human EGF, PDGF (beta form), or IGF-1 receptors. The assay relied upon the rapid activation of RTKs upon addition of ligand (Figure 4.4). In this case, the receptor complex forms an

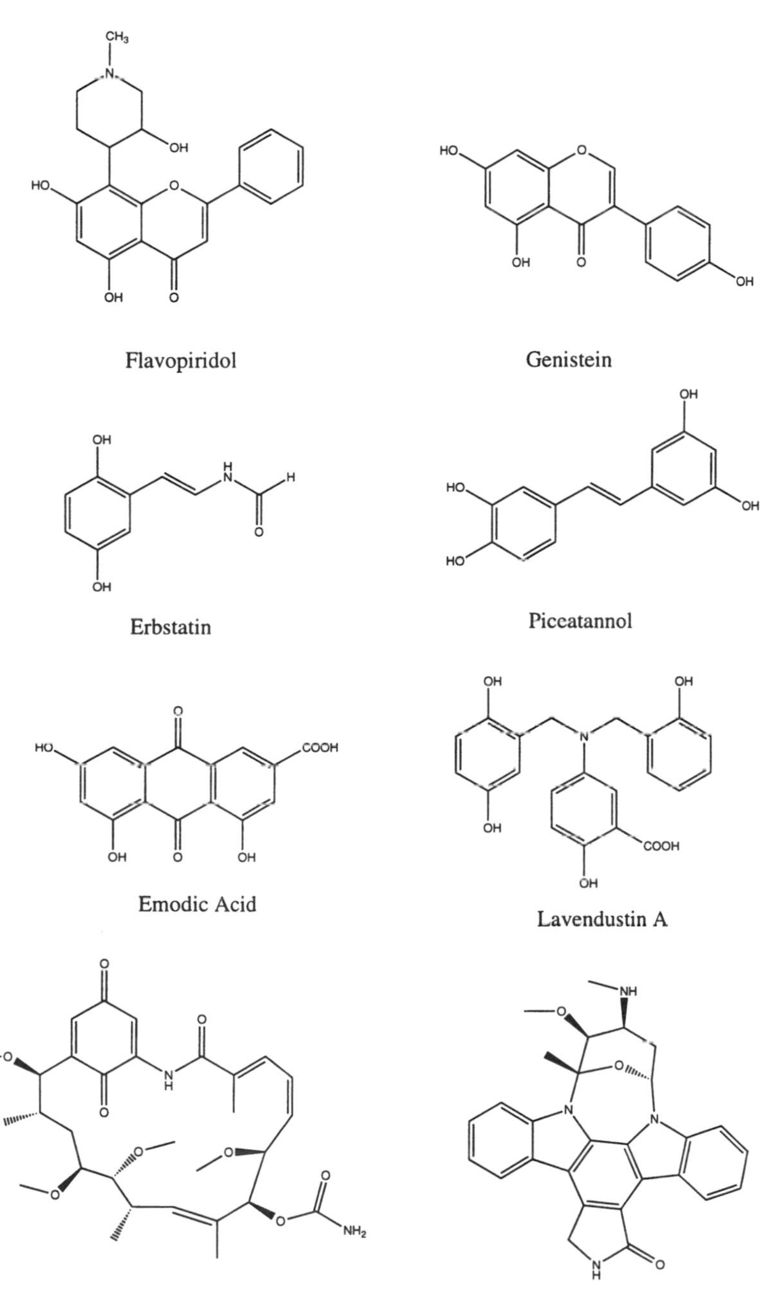

Figure 4.2 Kinase inhibitors derived from natural sources. Chemical structures corresponding to compounds derived from microbial extracts are shown. These compounds have been shown previously to exhibit kinase inhibitory activities.

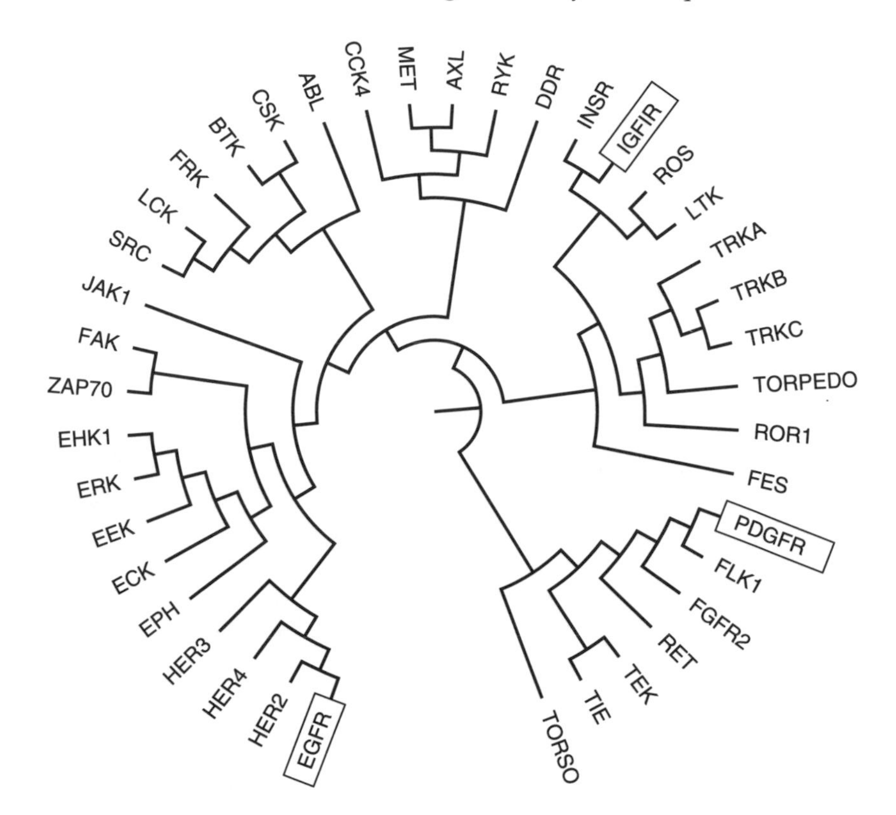

Figure 4.3 Comparison of tyrosine kinases. This dendrogram illustrates a comparison of the primary amino acid sequences associated with the catalytic domains for various tyrosine kinases, an analysis that leads to grouping of tyrosine kinases into sub-families. The EGF, IGF-1 and PDGF receptor tyrosine kinases are boxed, and represent three distinct tyrosine kinase subfamilies based upon the analysis of the catalytic core of these enzymes.

active homo-dimer resulting in ligand dependent phosphorylation of intracellular tyrosines on the receptor. Typically, extracts were resuspended in dimethyl sulfoxide at a concentration of 10 mg/ml and frozen in replicate 96-well microtitre plates until use. Extracts were diluted and incubated with cells for 2 h, after which the adherent cells were treated with either EGF, PDGF or IGF-1 to enable rapid activation of the receptor in each of the three cell lines used. After 10 min, the cells were lysed with a buffer containing detergent and *o*-vanadate (to inhibit cellular tyrosine phosphatases). As shown in Figure 4.5, the lysate was transferred to microtitre plates coated with monoclonal antibodies that are specific to one of the EGF, PDGF or IGF-1 receptors. The plates were washed to remove unbound lysate, and the amount of tyrosine phosphorylation associated

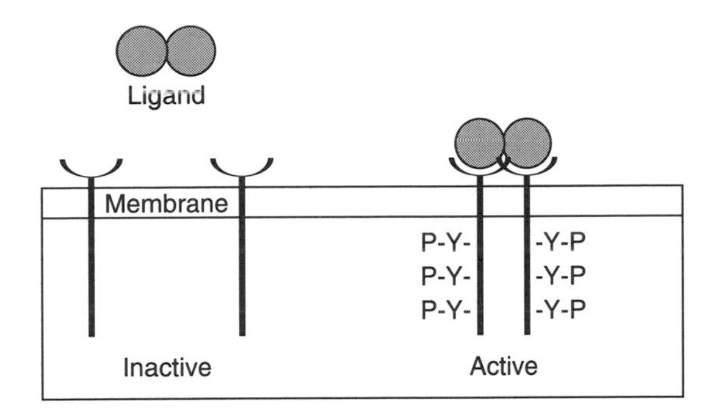

Figure 4.4 Ligand dependent dimerization and tyrosine phosphorylation of receptor tyrosine kinases. This schematic illustrates how the addition of ligand to inactive receptor kinases results in the dimerization and subsequent activation of the receptor. The activation of the receptor results in rapid phosphorylation of tyrosine residues that serve as recognition sites for signalling proteins resulting in cell growth, metabolism, differentiation, or survival.

with the receptor complex was quantitated using substrate based colorimetry following adsorption of horseradish peroxidase conjugated antiphosphotyrosine antibodies.

4.4 PRIMARY SCREENING RESULTS USING PLANT AND MICROBIAL EXTRACTS

In order to ensure good screen performance, cell based RTK assays have required pilot studies to determine the optimal dilution of extract. As shown in Figure 4.6, various dilutions of extracts resulted in different distribution profiles of inhibitory activities. This is illustrated using the PDGF receptor kinase assay and extracts derived from plant sources, where we found that a dilution factor of 1:250 was required in order to achieve the optimal distribution of activities. By contrast, we found that a dilution factor of 1:1000 was sufficient to achieve a similar distribution profile for the PDGF receptor tyrosine kinase using extracts derived from fungal sources. We feel that this difference most probably reflects the degree of chemical diversity associated with the membrane permeable fraction in the extract preparations. This diversity may be dependent upon the source material, solvent employed, and the purification procedures used to prepare the extracts. This screen analysis provided proof that our cell based assays exhibited good performance for both extract types.

After the pilot analysis, extract collections were then screened using the EGF, PDGF or IGF-1 receptor assays at the diluted concentration of extract deduced from the pilot studies. An analysis of inhibitory activities

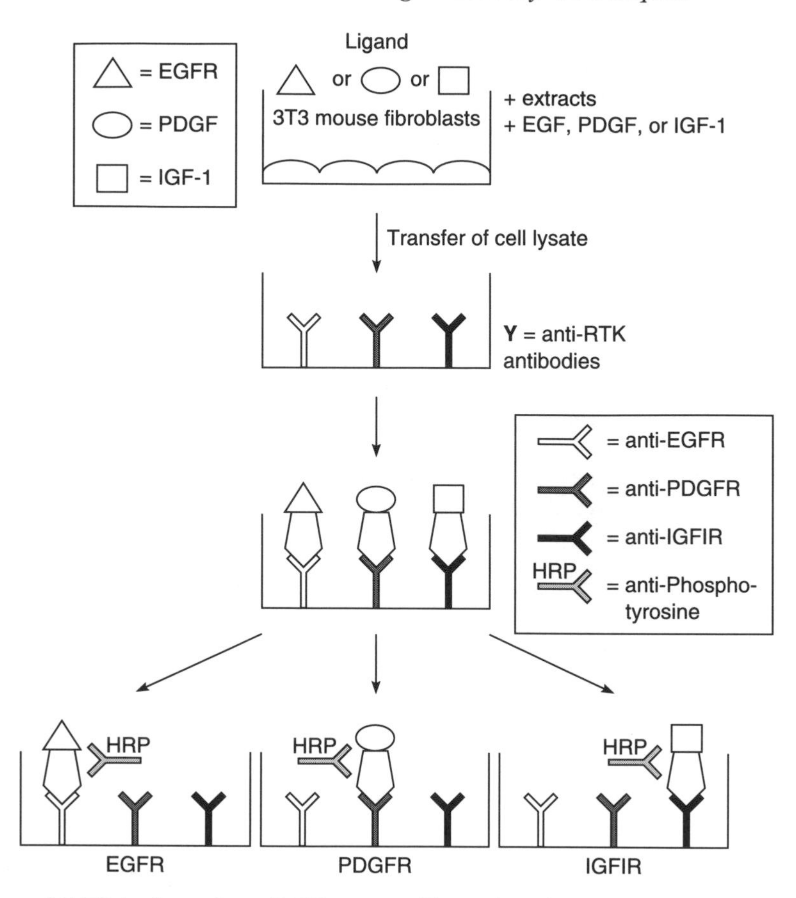

Figure 4.5 High throughput RTK assays, illustrating the procedure necessary when screening for inhibitors of RTKs using cell-based assays. Mouse fibroblasts that over-express the EGF, PDGF or IGF-1 receptors are plated into 96-well microtitre plates. Extracts are diluted and incubated with cells for 2 h, after which the cells are stimulated with EGF, or PDGF or IGF-1 for 10 min. Cells are then lysed with a detergent-based buffer and the lysates transferred to antibody-coated microtitre plates. Plates are washed and tyrosine phosphorylation associated with the EGF, PDGF or IGF-1 receptors are quantitated using colorimetry following adsorption of enzyme-linked anti-phosphotyrosine antibodies.

associated with each of the three assays indicated that extracts exhibiting specific inhibitory activities could be identified. This is illustrated in Table 4.1 using extracts derived from plant sources and screened at a dilution factor of 1:250. As noted, particular extract types show preferential inhibitory activites for the EGF, PDGF or IGF-1 receptor kinases. A similar analysis was performed for the microbial extracts and comparable results were obtained.

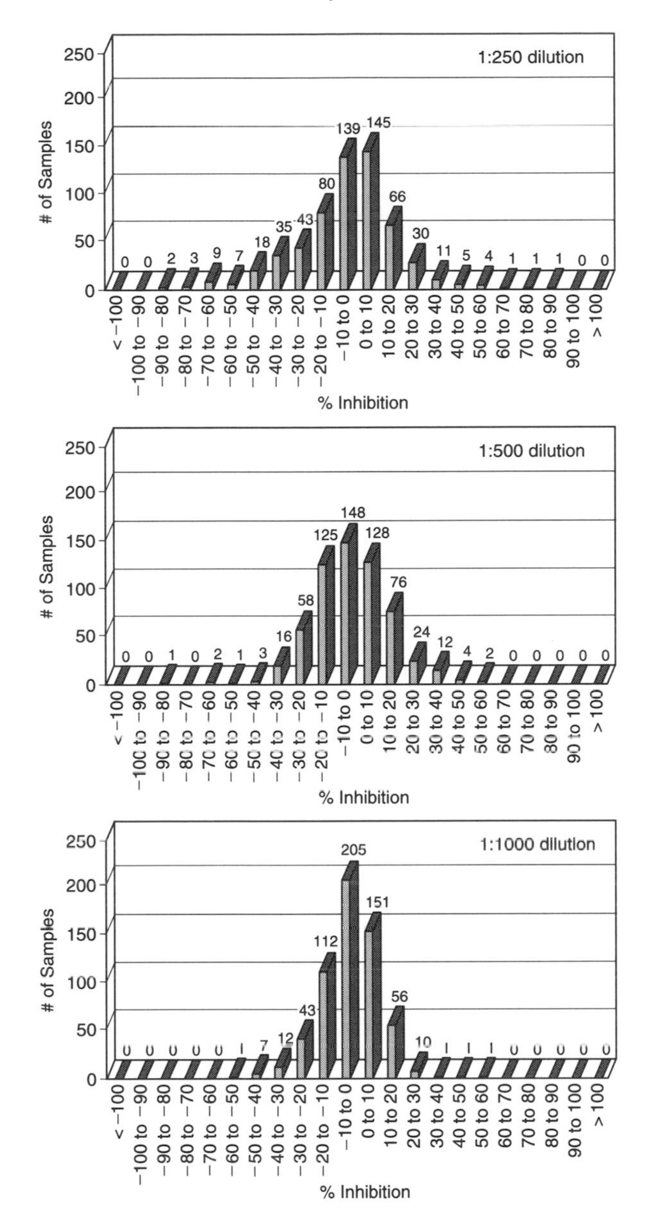

Figure 4.6 Distribution of cell-based PDGF kinase inhibitory activities associated with various dilutions of botanical extracts. Histograms corresponding to 1:250, 1:500 and 1:1000 dilutions of 10 mg/ml stocks of plant extracts illustrate the percentage of inhibitory activity associated with individual extracts. The percentage of inhibitory activity is calibrated in 10% increments and the number of samples associated with the inhibitory profile is denoted above each increment.

Table 4.1 Analysis of primary screening profiles using plant extracts for specific RTK inhibition

	Inhibition (%)		
Sample ID	*EGFR*	*PDGFR*	*IGF-IR*
pp-696193	39	86	23
pp-696332	−01	73	−03
pp-695811	−07	62	04
pp-695825	−03	59	08
pp-696225	23	59	02
pp-695793	62	58	35
pp-696025	15	54	03
pp-695993	65	45	55
pp-695969	−05	39	53
pp-696295	96	33	−21
pp-695833	07	28	63
pp-696188	51	25	03
pp-696277	06	21	57
pp-695908	09	15	50
pp-695942	18	09	67
pp-696033	09	09	86
pp-695917	07	06	55
pp-696204	64	06	−11
pp-696198	64	02	−00
pp-695794	55	01	−05
pp-696208	55	−03	08
pp-696217	56	−04	08
pp-695895	96	−05	15
pp-696158	63	−08	06
pp-696197	52	−11	32
pp-696168	56	−17	18
pp-696218	74	−21	18
pp-696178	62	−25	17
pp-696095	86	−69	−00

4.5 OPTIMIZATION OF SPECIFIC RTK INHIBITORY ACTIVITIES IN FUNGAL EXTRACTS USING FERMENTATION CONDITIONS

Generally, the level of metabolite production in microbial extracts is quite low (0.1–10 μg/ml of cultured broth). Fermentation optimization studies are essential for the efficient isolation of active metabolites. Although a few complex nitrogen and carbon sources can be used for the growth of microbes and production of metabolites, the use of media with different combinations of sources of carbon, nitrogen, and minerals has been shown to be useful to maximize metabolite production. The most common carbon sources for media optimization utilize glucose, glycerol, starch, dextrin, sucrose, maltose, and lactose at concentrations ranging from 2% to 5%. In the case of nitrogen sources, 1–3% soybean meal, cotton seed meal, wheat germ, peptone, sesame seed meal, and other additives are used. In addition, the use of fish meal or corn steep liquor at concentrations of 0.1–0.5% has been shown to increase the yield of active metabolites. Minerals such as calcium chloride and potassium-, sodium- and magnesium-containing compounds are required at concentrations over 0.02%. Salts of magnesium, iron, cobalt, and copper are required at concentrations around 0.002%. For instance, the addition of $FeSO_4{\cdot}7H_2O$ or $CoCl_2{\cdot}7H_2O$ sometimes can increase the metabolite yields appreciably. Using the principles cited above, we have been able to optimize the inhibitory activities for the EGF, PDGF and IGF-1 receptors. This is illustrated in Figure 4.7, where we show enrichment of PDGF, IGF-1 and EGF RTK inhibitory activities 3.4-, 4.3-, and 1.7-fold, respectively.

Prior to fractionation of the extract and elucidation of the active component, we sought to confirm the specific inhibitory activity using extracts prepared under conditions of optimized fermentation. In addition, we tested the extracts also using biochemical tyrosine kinase phosphorylation assays, which determine whether the inhibitory component is capable of blocking ATP dependent tyrosine phosphorylation using isolated receptors. In this case, receptors are localized using antibodies and the extract is diluted and added to isolated receptors followed by activation of the receptors with divalent cations and ATP. The identification of specific and potent inhibitions of the PDGF and IGF-1 receptor tyrosine kinases is shown in Table 4.2. In the case of the PDGF receptor, several optimized extracts from a single fungal strain show highly specific and potent inhibitory activity using both isolated receptors and intact cells when compared with both EGF and IGF-1 receptor kinases. By contrast, a different set of optimized extracts from another fungal strain showed potent and specific inhibition of IGF-1-dependent tyrosine kinase phosphoryation in cells but failed to show inhibitory activity using isolated IGF-1 receptors. These studies provide examples to justify the elucidation of active components for specific inhibitory activities for these two fungal strains.

4.6 CONCLUSIONS

We conclude that the use of cell-based RTK screening assays provides a means for the rapid identification of specific and potent activities from extracts derived from fungal and plant sources. The use of target-driven drug discovery principles and cross screening assays allows one to focus on extracts that have many of the features required for further consideration as *in vitro* leads. In addition, such principles allow attention to chemical purification and structure elucidation of the small subset of actives scored in primary screening. In the cases of the PDGF, EGF and IGF-1 receptors, the elucidation of active

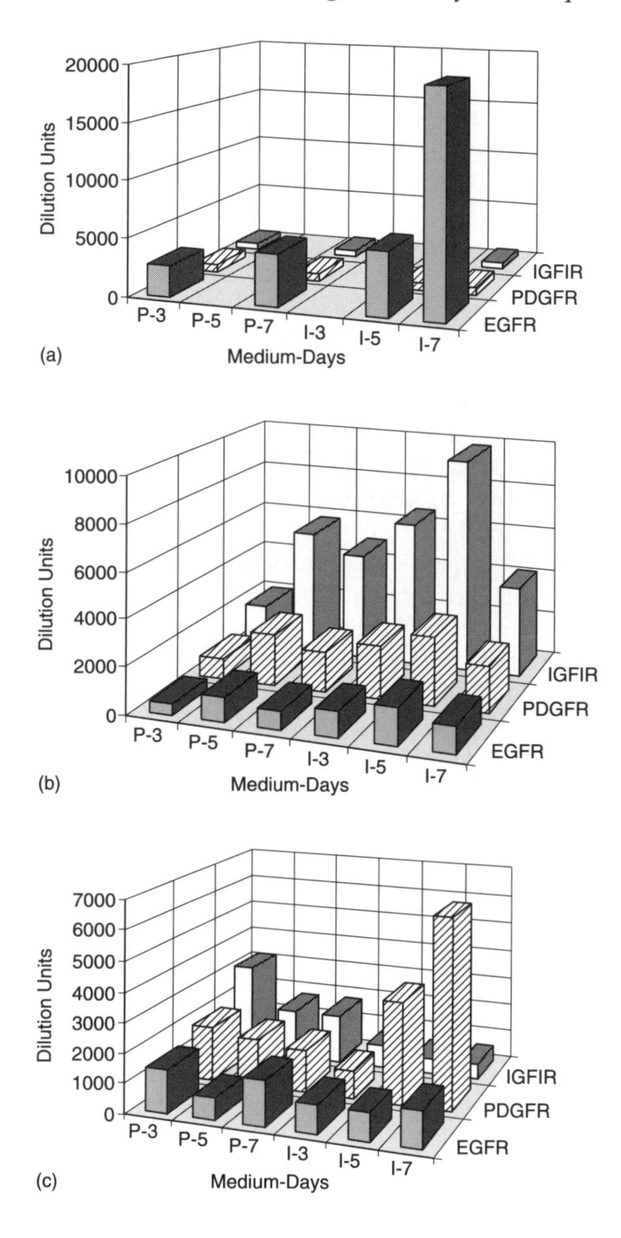

Figure 4.7 Optimization of RTK-specific inhibitory activities using various fermentation parameters. The graphs denote optimization of media and fermentation time for three actinomycetes exhibiting specific inhibitory activities for the EGF (a), PDGF (b) and IGF-1 (c) receptor tyrosine kinases. Extracts were prepared following 3, 5 or 7 days of fermentation in either medium P or medium I . Medium-days were plotted as a function of the assay tested (EGFR, PDGFR or IGFIR) and the dilution factor of the extract stock prepared in dimethyl sulfoxide at a concentration of 10 mg/ml.

Table 4.2 Analysis of optimized fungal extracts for relative specificity for inhibition of RTKs

Extract ID	*PDGFR cellular IC_{50} (dilution factor)*	*PDGFR biochemical IC_{50} (dilution factor)*	*EGFR cellular IC_{50} (dilution factor)*	*EGFR biochemical IC_{50} (dilution factor)*	*IGF-1R cellular IC_{50} (dilution factor)*	*IGF-1R biochemical IC_{50} (dilution factor)*
rok69F2	1,500,000	> 1,280,000	< 1,000	380	2,800	< 100
rok69F5	320,000	65,549	< 1,000	< 100	570	< 100
rok69F4	260,000	> 128,000	< 1,000	122	580	< 100
rok69F6	150,000	> 128,000	< 1,000	< 100	680	< 100
rok69F3	87,000	< 1,000	< 1,000	< 100	290	< 100
rok60E10	71,000	< 100	2,200	< 100	109,000	< 100
rok60E11	6,900	< 100	2,700	< 100	19,200	< 100
rok60E9	5,800	< 100	690	< 100	12,600	< 100
rok60E8	< 500	< 100	< 1,000	< 100	18,800	< 100
rok81D7	< 500	< 100	< 1,000	< 100	5,700	< 100
rok81D9	< 500	< 100	< 1,000	< 100	7,400	< 100

components that show specific inhibitory activities on these receptors represent *in vitro* leads for related RTK targets such as the Flk-1, HER2(neu), FGF receptors and other TK targets. These principles clearly are appropriate for other kinase targets of the tyrosine kinase or serine/threonine kinase variety, although the implementation of cell based assays may require some effort to engineer cell lines using chimeric receptors or inducible kinase activities. We feel that this drug discovery model has special significance using extracts derived from natural sources or mixtures of synthetic chemicals where active compounds may be under-represented or masked by unwanted activities.

ACKNOWLEDGEMENTS

We would like to acknowledge the contributions of Drs. Yuanying Wen, Xuemei Xie, Yangfang Chen, Zhihong Cheng, and Changxu Hu for provision of the plant extracts. We would also like to acknowledge Brad Smith, Brian Dowd, Kristen Cornuelle, Sarah Shimer, and Steve Vasile for performance and analysis of the screening assays. In addition, we would like to thank Martha Velarde and Holly Gregorio for preparation of the manuscript.

REFERENCES

Alvi, K.A., Nair, B., Gallo, C. and Baker, D. (1997) Screening of microbial extracts for tyrosine kinase inhibitors, *Japanese Journal of Antibiotics*, **50**, 264–6.

Chang, C.J. and Geahlen, R.L. (1992) Protein-tyrosine kinase inhibition: mechanism based discovery of antitumor agents, *Journal of Natural Products*, **55**, 1529–60.

Christian, M.C., Pluda, J.M, Ho, P.T., Arbuck, S.G., Murgo, A.J. and Sausville, E.A. (1997) Promising new agents under development by the Division of Cancer Treatment, Diagnosis and Centers of the National Cancer Institute, *Seminars in Oncology*, **24**, 219–40.

Fry, D.W., Kraker, A.J., McMichael, A., Ambroso, L.A., Nelson, J.M., Leopold, W.R., Connors, R.W. and Bridges, A.J. (1994) A specific inhibitor of the epidermal growth factor receptor tyrosine kinase, *Science*, **265**, 1093–5.

Groundwater, P.W., Solomons, K.R., Drewe, J.A. and Munawar, M.A. (1996) Protein tyrosine kinase inhibitors, *Progress in Medicinal Chemistry*, **33**, 233–9.

Levitzki, A. and Gazit, A. (1995) Tyrosine kinase inhibition: an approach to drug development, *Science*, **267**, 1782–8.

Plowman, G.D., Ullrich, A. and Shawver,L. K. (1994) Receptor tyrosine kinases as targets for drug intervention, *Drug News and Perspectives*, **7**, 334–9.

Powis, G., Bonjouklian, R., Berggren, M.M. , Gallegos, A., Abraham, R., Ashendel, C., Zalkow, L,. Matter, W.F., Dodge, J. and Grindley, G. (1994) Wortmannin, a potent and selective inhibitor of phosphatidylinositol-3-kinase, *Cancer Research*, **54**, 2419–23.

Rasouly, D. and Lazarovici, P. (1994) Staurosporine induces tyrosine phosphorylation of a 145 kDa protein but does not activate gp140trk in PC12 cells, *European Journal of Pharmacology*, **269**, 255–64.

Slate, D.L., Lee, R.H., Rodriguez, J. and Crews, P. (1994) The marine natural product, halistanol trisulfate, inhibits pp60v-src protein tyrosine kinase activity, *Biochemical and Biophysical Research Communications*, **203**, 260–4.

Yokoyama, A., Okabe-Kado, J., Uechara, Y., Oki, T., Tomoyasu, S., Tsuruoka, N. and Honma, Y. (1996) Angelmicin B, a new inhibitor of oncogenic signal transduction, inhibits growth and induces myelomonocytic differentiation of human myeloid leukemia HL-60 cells, *Leukemia Research*, **20**, 491–7.

Zhang, L., Chang, C.J., Bacus, S.S. and Hung, M.C. (1995) Suppressed transformation and induced differentiation of HER-2/neu-overexpressing breast cancer cells by emodin, *Cancer Research*, **55**, 3890–6.

5

Plant Cell Culture: a Vehicle for Drug Discovery

Angela M. Stafford, Christopher J. Pazoles, Scott Siegel and Li-An Yeh

5.1 INTRODUCTION

Traditional and modern medicines draw heavily on the chemical richness and diversity of the plant kingdom. The ethnopharmacological use of plants is becoming widely known through the scientific literature; and as screening technology becomes more accessible, the active principles of medicines used traditionally for hundreds of years are under scrutiny. In the form of 'phyto-ceuticals', plant-derived antioxidants are finding widespread acceptance as preventative medicines.

That plants continue to yield novel chemical structures with properties which can be harnessed for the treatment of human disease is illustrated well in Cragg, Newman and Snader (1997), which presents an analysis of natural products, approved by the FDA or in development 1983–1994, with particular reference to anticancer and anti-infective agents. Around 62% of approved anticancer drugs and pre-NDA anticancer drugs are natural or natural-derived, and about 10% are plant derived. Of the 93 new approved anti-infectives, 7 are natural products of which one is plant-derived (artemisinin, an antimalarial) and 45 are semi-synthetic derivatives, reflecting the continuing emphasis on the development of microbially derived antibiotics based upon familiar structures such as the antifungal azoles.

Since it is estimated that only 5–15% of the higher plant kingdom has yet undergone investigation for content of biologically active compounds, the potential for novel drug discovery is high. It was this potential combined with the relevance of plant cell culture technology in the context of drug discovery, to be described below, which drove the start-up of Phytera in 1992.

Advances in Drug Discovery Techniques. Edited by Alan L. Harvey © 1998 John Wiley & Sons Ltd.
ISBN 0 471 97509 5

5.2 PLANT CELL CULTURE TECHNOLOGY

The first successful attempts to culture isolated cells from plants were made by the eminent German botanist, Gottlieb Haberlandt (1902). He observed growth over a period of several weeks, but not cell division, in palisade cells. During the 1930's the role of B vitamins and auxins in plant growth was discovered, leading to the development of conditions for continuous growth and division in cultures derived from carrot root cambium and tobacco tumour tissue. White, Gautheret and Nobecourt are accredited with laying the foundations of plant tissue culture (Bhojwani and Razdan, 1996).

Interest in using plant cultures as an alternative to the whole plant for phytochemical production was first apparent in the 1950's when the Pfizer company filed a US patent in this area (Routier and Nickell, 1956). Twenty years later a 20 000 litre reactor was being used to culture cell suspensions of tobacco (Noguchi *et al.*, 1977), and the development had commenced of culture conditions not only supporting sustained growth on a large scale, but also supporting or inducing the production of valuable compounds. Significant progress was made in this area, for example with the development of highly effective indole alkaloid 'production media' as reported by Zenk (1977). The 1980's saw a high publication rate from scientists trying to enhance yields of numerous target compounds, first in undifferentiated plant cell cultures and later in differentiated 'hairy root' cultures induced by genetic transformation with *Agrobacterium rhizogenes*. The outcome of 10–15 years effort in this area was a large amount of data which can be summarized as follows: (i) cell suspension cultures of any target species can be developed with care; (ii) *commercially* viable levels of secondary products are achieved rarely in plant cell cultures (for some exceptions see below); (iii) certain classes of phytochemical appear to be associated with a level of differentiation and therefore do not accumulate to high levels in undifferentiated plant cell cultures; and (iv) undifferentiated plant cell cultures can produce novel chemical structures.

5.3 COMMERCIAL PRODUCTION SYSTEMS

Capital and running costs for large scale plant tissue culture systems are high, which discounts the production of anything other than very high value substances via this method. It is because so many plant secondary products can be defined as 'high value', usually on the basis of their pharmaceutical application, that the technological advances of the last 20 years in plant cell culture took place at all. However, the relatively small number of processes which reached commercialization is disappointing given the effort which went into research and development in the 1980's.

Mitsui Petrochemical Co. of Japan were the first to take a plant cell culture process to commercialization (Fujita, 1985). Their product, shikonin, was a kimono dye which could also be incorporated in cosmetics. The producing cell line had been selected rigorously to maximize shikonin content, and under the appropriate culture conditions yields of up to 15% of the dry biomass could be achieved, compared with only 2% in roots of the whole plant. Market size and breadth of application were the limiting factors in the relative success of this process. Although shikonin also has medicinal properties (anti-inflammatory, antibacterial), it has never been marketed widely as a pharmaceutical product.

Much more recently, Phyton Inc. in collaboration with Bristol Myers Squibb have developed a plant cell culture process for the production of paclitaxel. Phyton are growing cultures of *Taxus* spp. in a facility in Ahrensburg, Germany where volumes of up to 75 m^3 are possible. Commercialization of the cell culture route has yet to be realized, and its importance relative to alternative production methods such as semi-synthesis from the biosynthetic intermediate 10-deacetyl baccatin III has yet to be proven.

5.4 PHYTOCHEMICALS FROM PLANT CELL CULTURE

Compounds detected in undifferentiated cultures exhibit structural diversity (Table 5.1). While some products accumulate to very high levels in cultures of the appropriate plant species, e.g., shikonin (naphthaquinone), rosmarinic acid (phenylpropanoid), ginsenosides (triterpenoid saponins) and berberine

Table 5.1 Examples of secondary metabolites produced in plant cell suspension cultures (Mersinger, Dornauer and Reinhard, 1988; Fowler and Stafford, 1992; Van Hengel *et al.*, 1992; Stafford and Pazoles, 1997)

Chemical class	*Compound name*	*Plant species*
Phenolics		
Lignan	podophyllotoxin	*Podophyllum hexandrum*
Anthraquinone	sennosides A & B	*Rheum palmatum*
Naphthoquinone	shikonin	*Lithospermum erythrorhizon*
Flavonoid	anthocyanins	*Vitis vinifera*
Terpenoid		
Iridoid	valepotriates	*Valeriana officinalis*
Steroid	diosgenin	*Dioscorea deltoidea*
Diterpene	paclitaxel	*Taxus* spp.
Diterpene	forskolin	*Coleus forskohlii*
Alkaloid		
Indole alkaloid	ajmalicine	*Catharanthus roseus*
Isoquinoline alkaloid	berberine	*Coptis japonica*
Betalain alkaloid	betaxanthins	*Beta vulgaris*
Quinoline alkaloid	camptothecin	*Camptotheca acuminata*

(isoquinoline alkaloid), certain types of secondary metabolite can be found in trace qualities only. Morphine and vinblastine are two of the more important compounds which have yet to be reported reliably in a plant cell culture despite enormous scientific effort. In such cases it can be concluded only that some level of cellular or subcellular differentiation is essential for the product to be synthesized or accumulated to detectable levels.

However, this partial loss of metabolic capability is offset by new potential. The process of dedifferentiation appears to allow new biosynthetic routes to operate, as a result of which many novel structures have been reported in cell cultures which have not been observed in the corresponding intact plant. A review by Ruyter and Stockigt (1989) describes 85 novel structures from plant tissue cultures representing alkaloids, terpenoids, quinones and phenylpropanoids, including some novel skeletons. Since this review, the list has continued to grow (Table 5.2) and, given the very recent substantial effort in the *Taxus* culture arena, it is not surprising that some of the most recent additions are taxoid compounds (Ma *et al.*, 1994).

5.5 SCREENING PLANT CELL CULTURES: EARLY PROGRAMMES

For most of the novel structures isolated from plant tissue cultures no biological activity data have been reported. Some of the reported active novel structures resulted from a screening programme undertaken by the

Table 5.2 Novel structures reported in plant cell cultures (Ruyter and Stockigt, 1989; Ma *et al.*, 1994; Ishiguro *et al.*, 1995)

Compound (class of compound)	*Species*	*Family*
Phenolic		
Rutarensin (bis-coumarin)	*Ruta chalepensis*	Rutaceae
Paxanthone B (xanthone)	*Hypericum patulum*	Guttiferae
Licodione (retrochalcone)	*Glycyrrhiza echinata*	Leguminosae
Echinofuran B (benzoquinone)	*Lithospermum erythrorhizon*	Boraginaceae
Terpenoid		
Debneyol derivatives (sesquiterpenes)	*Nicotiana tabacum*	Solanaceae
Valtrate derivatives (iridoids)	*Valeriana wallichii*	Valerianaceae
Taxoid derivatives (diterpene)	*Taxus baccata*	Taxaceae
Alkaloid		
Epchrosin (indole alkaloid)	*Ochrosia elliptica*	Apocynaceae
Ajmalicine derivatives (indole alkaloid)	*Catharanthus roseus*	Apocynaceae
Voafrine A and B (dimeric indole alkaloids)	*Voacanga africana*	Apocynaceae
Gluco jatrorrhizine (berberine alkaloid)	*Berberis stolonifera*	Berberidaceae

Nattermann Company in Cologne (Kesselring, 1985; Schripsema *et al.*, 1996). In the course of this programme, callus cultures were established, then extracted and tested in a number of screens. From the ~10% of callus cultures exhibiting an activity, liquid suspensions were initiated and scaled up in order to produce sufficient biomass for identification of the active compounds. The overall duration from whole plant to isolated compound was estimated to be 2–4 years. Forty substances with pharmacological activity were isolated, of which 26 were anti-inflammatory. These represented a wide range of chemical structures and 7 were novel. Some results of this work were published during the 1980's (Arens *et al.*, 1982, 1985; Nattermann, 1986) and although none of the compounds appears to have been progressed to the clinic, the outcome illustrates the fact that undifferentiated plant cultures can yield novel chemical structures with interesting biological activities.

A much smaller survey of just 50 cell cultures was carried out by Berlin *et al.* (1984), specifically to determine cytotoxic activity. Their conclusion was that no novel cytotoxic activities could be found in extracts of these cultures and that when culture activity was found, it was generally seen in the source plant at a higher level.

5.6 THE PHYTERA APPROACH TO LEAD DISCOVERY: ExPAND™

Phytera has chosen to base its lead discovery effort at least in part on plant cell cultures, generated in Sheffield and more recently in Copenhagen. Culture extraction, biological screening, and natural product and synthetic chemistry are based in the Company's US headquarters where a marine microorganism collection also is being generated and exploited.

Phytera's extract bank fuels its own screens and those of its partners, and provides an additional reason for focusing on plant cultures rather than whole plants for lead structures: ensured resupply. The cultures are a renewable resource which can be used to generate fresh extracts on demand.

ExPAND™ technology (expanded phytochemistry aimed at novel discovery) comprises 'diversity-directed' plant sourcing, plant cell culture and plant cell culture manipulation, the combined strategy leading to culture extracts with a greatly expanded phytochemical repertoire (Figure 5.1). The core rationale behind the focus on plant cell cultures is their 'pluripotential'; providing theoretical access to the entire genome and therefore biosynthetic capacity. The involvement of culture manipulation makes the Phytera approach fundamentally different from that used in the earlier published screening programmes using plant cell cultures or whole plants. Rather then taking 'phytochemical snapshots', the Phytera strategy builds an 'album' for each species, thereby providing the screening process with greater opportunity for the discovery of new actives.

All cultures are taken to the liquid suspension stage before primary screening (Figure 5.2). Prior to harvest, replicate samples of cultures are exposed to a

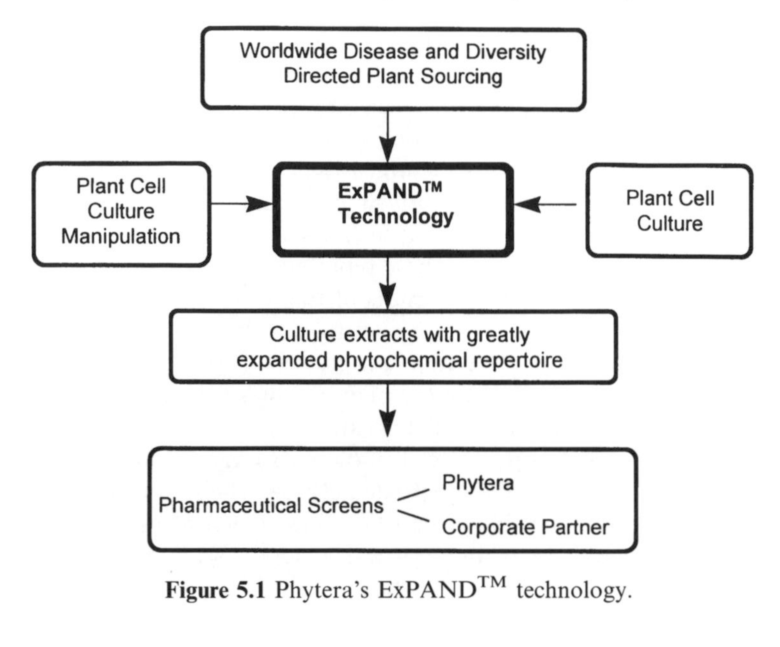

Figure 5.1 Phytera's ExPAND™ technology.

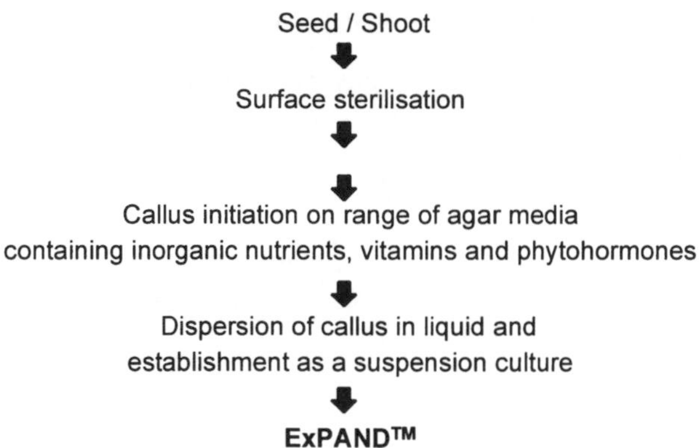

Figure 5.2 Plant cell culture strategy.

battery of manipulations designed to maximize phytochemical diversity per species. These manipulations include variation in medium composition, 'elicitation' of defence responses in plant cell cultures, and proprietary gene derepression strategies. Precedents for some of these manipulations can be found in the literature; for example, alterations in the hormone content and sugar composition of the medium contributed to the indole alkaloid enhancing effect of the production medium reported by Zenk *et al.* (1977). Subsequent studies have built on this work and shown that the synthetic auxin 2,4-D, widely used in plant tissue culture to stimulate cell growth, also suppresses secondary metabolism.

A large proportion of phytochemistry is represented by inducible small molecular weight defence compounds, or phytoalexins, the most widely studied of these being phenylpropanoid and sesquiterpenoid structures. The production of such compounds can be induced in undifferentiated cell cultures using a variety of agents to mimic 'attack'; for example, whole autoclaved preparations of fungal cultures, or fractions of fungal cell walls. Similar phytochemical effects can be induced by targeting further downstream in the plant response to infection or wounding by adding jasmonates (Wasternack and Parthier, 1997). The significance of the role of jasmonates is not understood clearly, but they appear to act as 'signal transduction' intermediates in a cascade of biochemical events not dissimilar to that leading to the production of prostaglandins in animals.

A fuller understanding of human and pathogen metabolism can provide new methods for altering plant metabolism in culture. For instance, many human drugs are biosynthetic inhibitors which might be expected to inhibit the analogous enzymes in plants. Selective inhibitors of sterol biosynthesis have been shown to alter the spectrum of sterols in plant cell cultures, e.g., in cell suspensions of fenugreek (Cerdon *et al.*, 1995) when treated with the triazole fungicide diniconazole. HMG CoA reductase inhibitors developed as hypolipodaemic drugs such as mevinolin might be expected to reduce flux down the entire terpenoid pathway. However, as alternative biosynthetic routes to the terpenoids may exist in plants, the effects of such effectors cannot be predicted with certainty. Thus, the use of enzyme inhibitors and precursors to channel biosynthetic pathways must still be regarded as semi-empirical.

Harvested cultures are processed by generating extracts of differing polarities from biomass and media; these samples then pass into a range of screens in-house based upon antifungal, antiviral and antibacterial targets (both whole cell and molecular).

An analysis of $>150\,000$ screening events of our culture extracts over a number of (mainly) anti-infective screens shows that each manipulation adds value in terms of 'hit rate' (on average $<1\%$ of all extracts). Analysis of confirmed hits, and of subsequent leads, shows that 90% of these derive from manipulated cultures (as opposed to those cultivated on standard growth media) and that about 40% are exclusive to a single manipulation. Very few confirmed hits are seen in more than three manipulations simultaneously.

Another important result is the discovery of novel chemistry as the basis of a proportion of these leads; furthermore, in some instances when extensive analysis of whole plant tissue also has been performed, the unique active has not been detected, demonstrating that secondary metabolic routes in cell cultures do indeed differ from those of their source plant; in fact, more often than not, the cell culture chemical fingerprints are seen to be more complex.

Our strategy is yielding new biologically active structures of which, because of our screening focus in-house, we are presently detecting only those with anti-infective properties. With collaborative partners we are targeting other therapeutic areas in order to exploit further, as yet undiscovered, culture chemistry.

The discovery of a particularly interesting lead from *in vitro* studies may trigger a decision to enter the candidate for pre-clinical testing in animal models. At the same time, knowledge of the chemical structure of the active principle can provide a platform for structural analoguing, which with today's rapidly emerging combinatorial chemistry strategies can result in many thousands of related structures in a very short space of time. Given the pressure that this imposes on screening capacity, it is obvious that such strategies are adopted only for candidate drugs of considerable interest.

5.7 CASE STUDY: THE DEVELOPMENT OF ANTIFUNGAL DRUG CANDIDATES

5.7.1 The need for new antifungal therapies

While the worldwide incidence of serious fungal infections is increasing, effective therapies are limited. Amphotericin B, a polyene antifungal with toxic side-effects, and the azole antifungals, notably fluconazole, are presently the major treatments for systemic fungal infections (Graybill, 1996). The emergence of azole-resistant fungi adds urgency to the search for new antifungal drugs, particularly those which may have a different site of action. New azoles are being developed by a number of companies, and other classes with potential include echinocandins, nikkomycins, and very recently reported sordarins, a series under development at Glaxo.

In general, any new antifungal agents should have an excellent safety profile and a good spectrum of activity in order to be competitive with existing therapies.

5.7.2 Sunillin: *in vitro* and *in vivo* studies

Sunillin is the name given to a drug candidate discovered in and isolated from a plant cell culture manipulated using Phytera's ExPAND^{TM} technology. This candidate cannot be detected in the native plant, only in manipulated cultures (Figure 5.3), and has a molecular weight of 354 Da.

In vitro studies have demonstrated that sunillin has broad spectrum antifungal activity against several strains of *Candida* spp. including some fluconazole resistant isolates, *Crytococcus neoformans* and *Aspergillus fumigatis*. Sunillin has also been tested *in vitro* in combination with three other classes of antifungal drug (two approved, one in clinical development); the results indicate strong synergy, which is not unexpected as sunillin appears to have an entirely different mode of action from that of other drugs, and does not exert its primary effect on fungal cell wall synthesis.

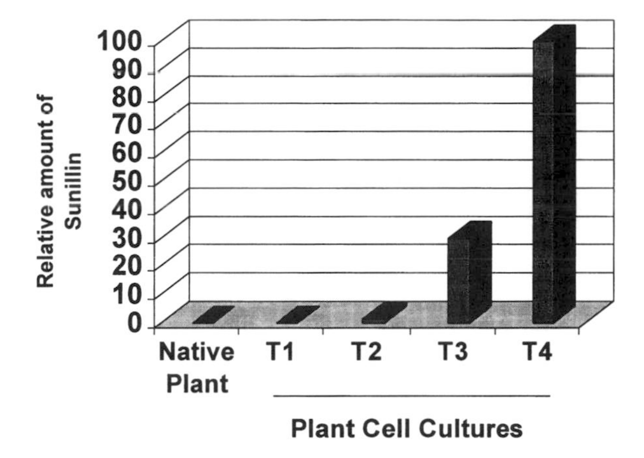

Figure 5.3 Sunillin is produced in manipulated cultures but not in the native plant (manipulations = T1 . . . T4).

A murine model of lethal disseminated candidiasis has been used to assess the *in vivo* efficacy of sunillin, administered either intraperitoneally or orally, on a daily basis for 10 days, 24 hours after infecting the mice. In these studies, sunillin was found to protect via both routes of administration, and its activity compared very favourably with that of fluconazole given twice daily. Vehicle-treated mice died with 16 days of infection. When organ fungal burdens were determined in surviving mice, sunillin was substantially more effective than fluconazole.

The primary *in vivo* studies were performed using a fluconazole-sensitive strain of *Candida albicans*. When similar experiments were performed with a fluconazole-resistant strain, the results were striking; fluconazole administered orally was poorly effective whereas sunillin protected to the same extent as seen in the earlier trials (Figure 5.4). The fungal burden in kidney, liver and spleen was reduced almost to zero by day 30 by oral dosing with sunillin (Figure 5.5).

Further *in vivo* studies are in progress to evaluate the spectrum of activity more fully, and the efficacy of sunillin in combination therapy. Single and multiple oral dosing regimes with up to ×10 protective dose so far have shown no observable toxic effects for sunillin.

5.7.3 Status

Sunillin is now being produced by a synthetic route and the process is being scaled up to kilogram level in external GMP facilities. Further toxicology trials are planned and pharmacokinetic studies are in progress. The development of this candidate is therefore well underway, and other related candidates are following. So far, more than 4000 sunillin analogues have been prepared at

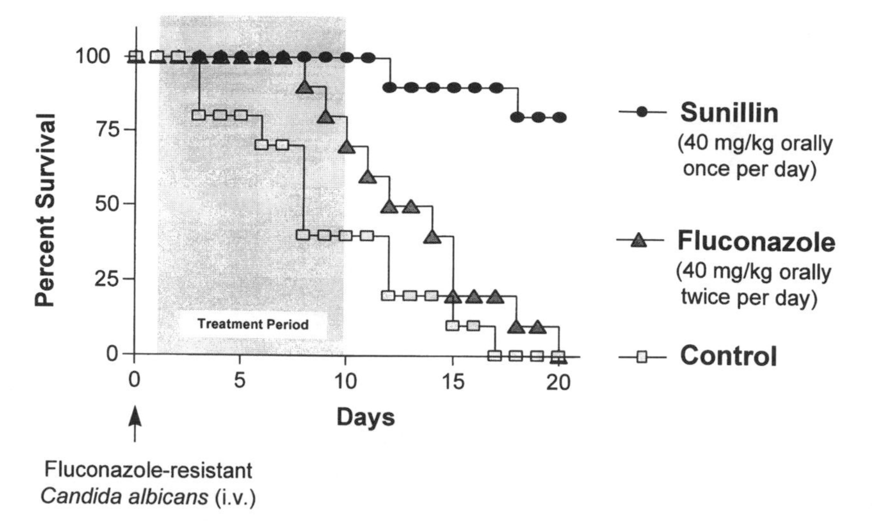

Figure 5.4 Sunillin is effective *in vivo* against a fluconazole-resistant strain of *Candida albicans* (UTR-14).

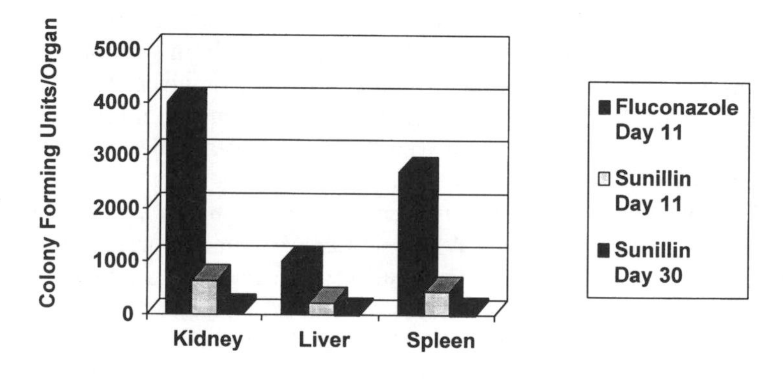

Figure 5.5 Sunillin decreases the organ fungal burden in a murine model of fluconazole-resistant candidiasis.

Phytera via combinatorial chemistry. Screening of these compounds has resulted in the identification of another 10 lead compounds, some displaying distinctly improved potency *in vitro* against filamentous fungi such as *Aspergillus fumigatis*.

So far, the Phytera strategy has yielded novel chemistry and a series of potent antifungals with in all likelihood a new mode of action. Other lead structures, some novel, with activities against other therapeutic targets (e.g., herpes, CMV, hepatitis, and antibiotic-resistant bacteria) are poised to follow a similar development route. The future success of our drug discovery effort continues to rely upon the introduction and expansion of chemical diversity in our plant culture bank, discriminating chemical extraction methods and carefully selected screening targets. The ability to reaccess chemistry via stored plant cultures provides an important advantage over more conventional phytochemical screening approaches. With the addition of more partnerships to our portfolio, we expect to see many more drug candidates in a variety of therapeutic areas advancing towards the clinic in the near future.

ACKNOWLEDGEMENTS

Thanks go to numerous colleagues at Phytera who have contributed to the studies reported above, including Mike August, Alex Chu, Catherine Heintz, Jim McAlpine and Margaret Spencer.

REFERENCES

Arens, H., Borbe, H.O., Ulbrich, B. and Stöckigt, J. (1982) Detection of pericine, a new CNS-active indole alkaloid from *Picralima nitida* cell suspension culture by opiate receptor binding studies, *Planta Medica*, **46**, 210.

Arens, H., Fischer, H., Leyck, S., Roemer, A. and Ulbrich, B. (1985) Anti-inflammatory compounds from *Plagiorhegma dubium* cell culture, *Planta Medica*, **49**, 52–6.

Berlin, J., Forche, E., Sasse, F. and Schairer, H.U. (1984) Comparison of cytotoxic activities of plants and their cultures, *Farmacologische Tijdschrift Belgium*, **61**, 330.

Bhojwani, S.S. and Razdan, M.K. (1996) Introductory history, in *Studies in Plant Science*, Vol. 5, *Plant Tissue Culture: Theory and Practice*, Revised Edn, Elsevier, Amsterdam, pp. 1–18.

Cerdon, C., Rahier, A., Taton, M. and Sauvaire, Y. (1995) Effect of diniconazole on sterol composition of roots and cell suspension cultures of fenugreek, *Phytochemistry*, **39**, 883–93.

Cragg, G.M., Newman, D.J. and Snader, K.M. (1997) Natural products in drug discovery and development, *Journal of Natural Products*, **60**, 52–60.

Fowler, M.W. and Stafford, A.M. (1992) Plant cell culture, process systems and product synthesis, in *Plant Biotechnology – Comprehensive Biotechnology*, Second Supplement (eds M.W. Fowler, G.S. Warren and M. Moo-Young), Pergamon Press, Oxford, pp. 79–98.

Fujita, Y. (1985) Production of plant pigments by plant tissue and cell culture, *Journal of Synthetic Organic Chemistry*, **43**, 1003–12.

Graybill, J.R. (1996) The future of antifungal therapy, *Clinical Infectious Diseases*, **22** (Suppl. 2), S166–78.

Haberlandt, G. (1902) Kultinversuche mit isollerten Pflanzellen, *Sitzungsberichte der Akademie der Wissenschaften in Wien*, **111**, 62–9.

Ishiguro, K., Nakajima, M., Fukumoto, H. and Isoi, K. (1995) Co-occurrence of prenylated xanthones and their cyclization products in cell suspension cultures of *Hypericum patulum*, *Phytochemistry*, **38**, 867–9.

Kesselring, K. (1985) Pflanzenzellkulturen (PZK) zur Auffindung neuer, therapeutische relevanter Naturstoffe und deren Gewinnung durch Fermentationsprozesse, in *Pflanzliche Zellkulturen*, Bundesministerium fuer Forschung und Technologie, Bonn, p. 111.

Ma, W., Park, G.L., Gomez, G.A., Nieder, M.H., Adams, T.L., Aynsley, J.S., Sahai, O.P., Smith, R.J., Stahlhut, R.W. and Hylands, P.J. (1994) New bioactive taxoids from cell cultures of *Taxus baccata*, *Journal of Natural Products*, **57**, 116.

Mersinger, R., Dornauer, H. and Reinhard, E. (1988) Formation of forskolin by suspension cultures of *Coleus forskohlii*, *Planta Medica*, **54**, 200–3.

Nattermann & Cie GmbH (1986) Dehydrodiconiferyl alcohol glucosides production from leaf cell cultures of *Plagiorhegma dubium*; new anti-inflammatory, German Patents DE 3438886: 24.10.84; DE438886: 24.04.86.

Noguchi, M., Matsumoto, T., Hirata, Y., Yansamoto, K., Karsuyama, A., Kato, A., Azechi, S. and Kato, K. (1977) Improvement of growth rates of plant cell cultures, in *Plant Tissue Culture and its Biotechnological Application* (eds W. Barz, E. Reinhard and M.H. Zenk), Springer-Verlag, Heidelberg, pp. 85–94.

Routier, J.B. and Nickell, L.G. (1956) Cultivation of plant tissue, US Patent 2, 747, 334.

Ruyter, C.M. and Stockigt, J. (1989) Novel natural products from plant cell and tissue culture – an update, *GIT Fachzeitschrift fur die Laboratorie*, **4**, 283.

Schripsema, J., Fung, S.Y. and Verpoorte, R. (1996) Screening of plant cell cultures for new industrially interesting compounds, in *Plant Cell Culture Secondary Metabolism: Toward Industrial Application* (eds F. Di Cosmo and M. Misawa), CRC Press, Boca Raton, pp. 1–10.

Stafford, A.M. and Pazoles, C.J. (1997) Harnessing phytochemical diversity for drug discovery: the Phytera approach, in *Phytochemical Diversity, A Source of New Industrial Products* (eds S. Wrigley, M. Hayes, R. Thomas and E. Chrystal), The Royal Society of Chemistry, London, pp. 179–89.

Van Hengel, A.J., Harkes, M.P., Wichers, H.J., Hesselink, P.G.M. and Buitelaar, R.M. (1992) Characterisation of callus formation and camptothecin production by cell lines of *Camptotheca acuminata*, *Plant Cell, Tissue and Organ Culture*, **28**, 11–18.

Wasternack, C. and Parthier, B. (1997) Jasmonate-signalled plant gene expression, *Trends in Plant Science*, **2**, 285–323.

Zenk, M.H., El-Shagi, H., Arens, H., Stöckigt, J., Weiler, E.W. and Deus, B. (1977) Formation of the indole alkaloids serpentine and ajmalicine in cell suspension cultures of *Catharanthus roseus*, in *Plant Tissue Culture and its Biotechnological Application* (eds W. Barz, E. Reinhard and M.H. Zenk), Springer-Verlag, Berlin, pp. 27–32.

6

Pharmacy of the Deep – Marine Organisms as Sources of Anticancer Agents

Marcel Jaspars

6.1 INTRODUCTION

Recently, the hunt for new lead compounds, or pharmacophores, has become technology driven, and currently it is possible to test over a quarter of a million compounds in a single disease screen in one month using robotics, a number that is increasing rapidly (see Chapter 11 by Norrington). As a result, chemists have resorted to combinatorial chemistry to generate enough compounds to keep up with this demand (see Chapter 9 by Shuttleworth). Whether this highly automated strategy is successful is yet to be proved. In effect, combinatorial chemistry for lead generation is simulating the process of natural selection, by synthesizing many related compounds and selecting those with the most desirable properties. Nature has been carrying out this process for 3.8 billion years, and has generated some complex molecules with exquisite biological properties. It is widely believed that these compounds have been evolved in nature because of the selectional advantages obtained as a result of the interactions of these compounds with specific receptors in other organisms (Stone and Williams, 1992). This is evidenced by the fact that over 60% of currently approved drugs for the treatment of cancer are of natural origin, (Cragg, Newman and Snader, 1997). In this instance 'natural origin' is defined as natural products, derivatives of natural products or synthetic pharmaceuticals based on natural product models. Most of these natural products are derived from terrestrial plants and microorganisms (Aszalos 1981), and one problem is that the percentage of these showing positive activity in pre-screens is very low. Data from the National Cancer Institute's preclinical antitumour drug

Advances in Drug Discovery Techniques. Edited by Alan L. Harvey
ISBN 0 471 97509 5

discovery screen shows that of over 18 000 terrestrial plant extracts and over 8000 terrestrial microorganism extracts tested, only 0.4% showed significant selective cytotoxic activity (defined as an IC_{50} < 0.1 mM) (Garson, 1994). For this reason many investigators decided to start focusing on organisms from previously inaccessible niches. One vast habitat that was unexplored until recently is the marine environment. The development of the aqualung by Jacques Cousteau in the middle of the twentieth century made possible the collection of marine organisms from shallow waters. This allows effective collection to be carried out down to depths of about 40 m, and deeper collections can be performed now using remote operated vehicles. Since there was no scientific precedent or evidence from traditional medicine, most of the initial collections attempted to gather as diverse a selection of organisms as possible. Another development was necessary in order to speed up the discovery of natural products. Ready availability of 1-dimensional NMR techniques in the early 1970's and 2-dimensional NMR methods in the early 1980's accelerated the structural elucidation of complex natural products on increasingly smaller amounts (Crews, Rodriguez and Jaspars, 1998). Up to the end of 1996, about 10 000 compounds had been isolated from marine organisms (Blunt and Munro 1996), as compared with > 120 000 natural products reported from all sources (Buckingham, 1997). The distribution of these marine natural products between the different phyla is as shown in Figure 6.1. As can be seen, about a third of all reported marine natural products are derived from sponges (Porifera).

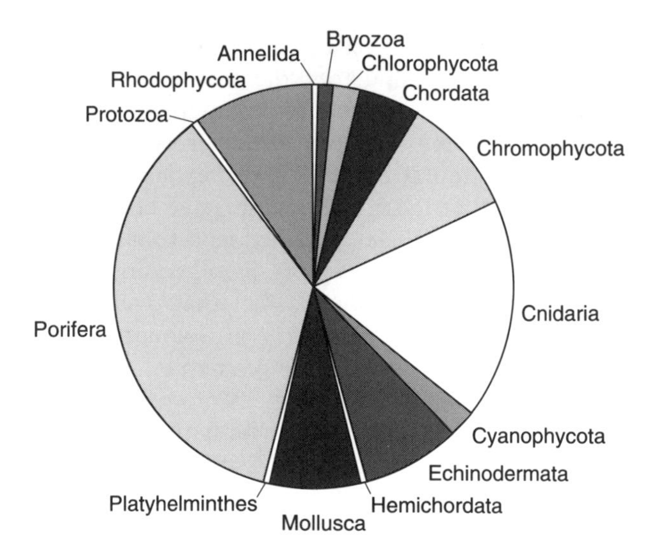

Figure 6.1 Percentage of marine natural products reported by phylum (Arthropoda, Brachiopoda, Euglenophycota, Nematoda all less than 1%).

The NCI's preclinical antitumour drug discovery screen shows that of over 6000 marine animal extracts tested, about 1.8% showed significant selective cytotoxic activity, whereas the figure for approximately 2000 marine plant extracts tested drops down to a disappointing 0.2% (Garson, 1994). Once we break this down into the phyla concerned an interesting picture emerges (Figure 6.2 (Garson, 1994)).

The two most prominent phyla are the Porifera (sponges) and Bryozoa, or more correctly Ectoprocta (seamats), and they show 'hit rates' of greater than 10%. The low percentage of compounds from bryozoans witnessed in Figure 6.1 is perhaps a manifestation of the difficulty in collecting bryozoans, and also because often they are confused with calcareous algae and stylasterine corals (Colin and Arneson, 1995). Sponges show perhaps a greater taxonomic diversity (estimates of 10 000 species or more are common), and are easier to collect, due to their larger size relative to bryozoans, and easier to identify to the genus level in the field. This then accounts for their greater popularity with marine natural product chemists. However, if one were to look simply for cytotoxicity

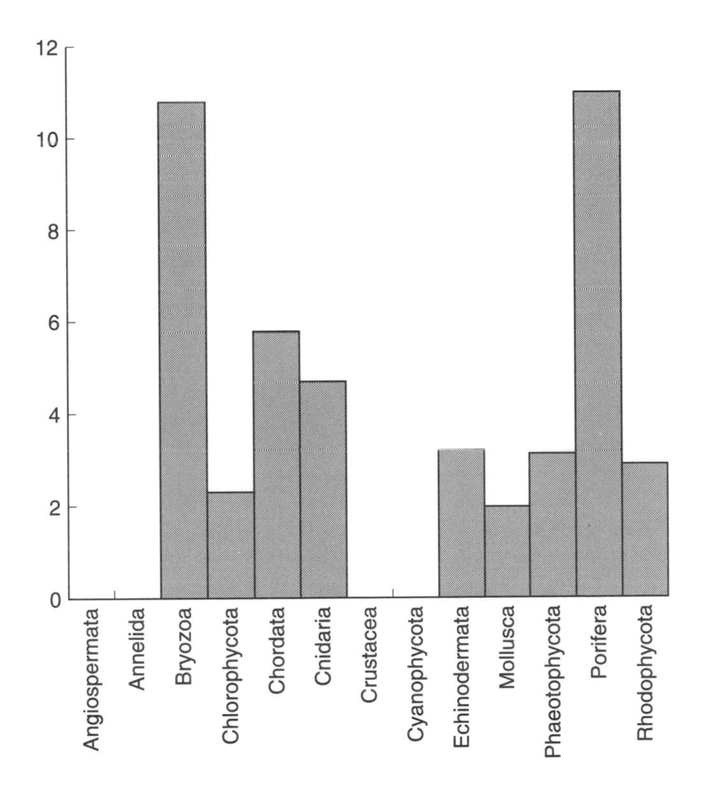

Figure 6.2 Percentage of organisms per phylum showing significant selective cytotoxic activity.

as reported in the primary literature, a different picture emerges (Figure 6.3). Using these data we should concentrate all our efforts on the platyhelminths, yet this has not occurred. It is difficult to compare quantitatively cytotoxicity data obtained using different cell lines, different protocols and in different laboratories. Also the extent of the cytotoxicity has not been taken into account. In one person's book a millimolar IC_{50} may count as 'cytotoxic' whereas another researcher may not report it until the IC_{50} is nanomolar or less. This shows the importance of obtaining consistent and comparable bioactivity data, such as provided by the NCI for Figure 6.2. Another factor in this case is the sample size, 14 of the 20 compounds isolated from platyhelminths show 'cytotoxicity' whereas 'only' 900 of the 4000 compounds isolated from sponges show cytotoxicity.

One assessment of the pharmaceutical potential of a compound is whether or not a patent has been filed for it. Over 200 patent applications were filed in the period 1969–1993 for marine natural products, of which roughly half fall into the antitumour/antiviral bracket. Of these patented compounds, 75 are

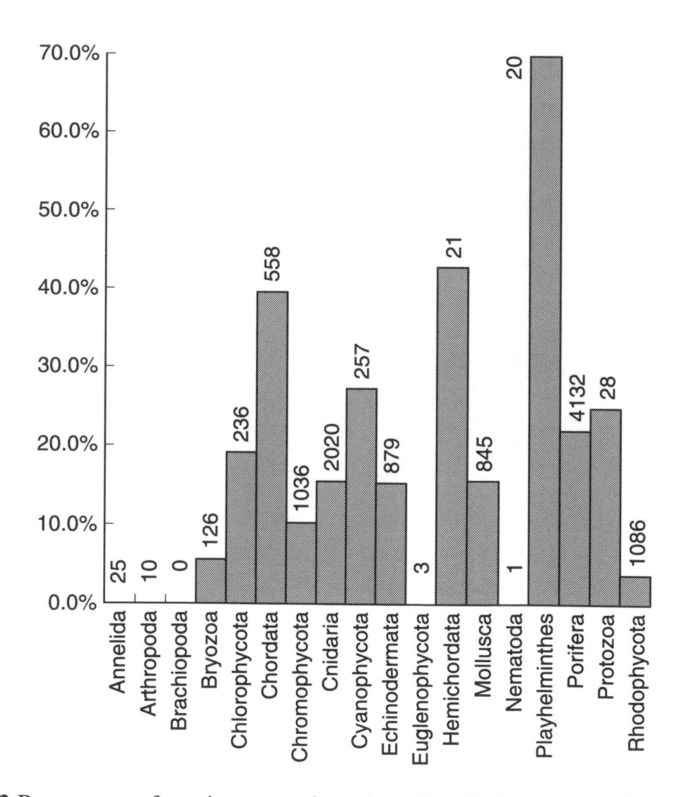

Figure 6.3 Percentage of marine natural products by phylum reported to possess cytotoxicity (total compounds reported given above column).

produced by sponges, 14 by chordates (ascidians), 10 by molluscs, 4 by hemichordates and 3 by bryozoans (Bongiorni and Pietra, 1996). Of course, the real measure of utility is whether any drug derived from marine sources has been approved for use. Of the 93 drugs currently available for the treatment of cancer, none is of marine origin (Cragg, Newman and Snader, 1997). However, of the 50 natural products which were the subject of a new drug application (NDA) for the treatment of cancer in the period 1989–1995, 9 were of marine origin, with compounds being derived from sponges (4), chordates (ascidians) (2), a mollusc (1), a bryozoan (1) and a red alga (1). These and other potential candidates have been reviewed in the past (Faulkner, 1993; Ireland *et al.*, 1993; Schmitz, Bowden and Toth, 1993; Munro *et al.*, 1994). The marine environment has placed a very different evolutionary pressure on marine organisms compared with terrestrial organisms, and this is evident from the extreme novelty of structure exhibited by marine natural products (see following sections). The 9 compounds that have been the subject to NDA's now follow, together with descriptions of their modes of action and the current status in clinical trials.

6.2 AEROPLYSININ

Aeroplysinin (**1**) was isolated first from a sponge collected in the Bay of Naples and identified initially as *Aplysina aerophoba* (Fattorusso, Minale and Sodano, 1970) sometimes also classified as *Verongia aerophoba* (Fattorusso, Minale and Sodano, 1972). One study suggested that this sponge should be assigned as *Verongia cavernicola* (all are sponges of order Verongida and family Aplysinidae) (Cimino *et al.*, 1983). The sponge is distributed widely in the Mediterranean, and has been found in the Atlantic near the Canary Islands. The compound is a modified bromotyrosine structure, a motif that appears repeatedly in marine natural products (Jaspars *et al.*, 1994). Aeroplysinin has been shown to be a defence metabolite for the sponge, and is released upon predation by the enzymatic breakdown of larger, less toxic secondary metabolites (Ebel *et al.*, 1997). Aeroplysinin has antibacterial properties, and also it

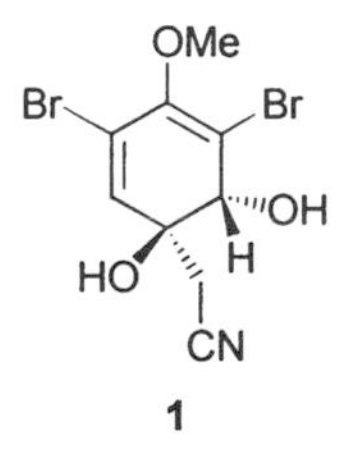

Structure 6.1

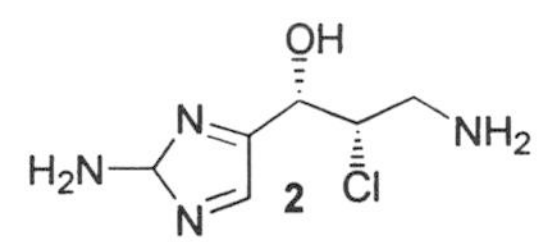

Structure 6.2

has been found to have activity against breast cancer, lymphoma, epithelioma cells *in vitro* (Kreuter *et al.*, 1989), and also HeLa cells (human cervix uteri tumour cell line) with an IC_{50} of 3.0 μM (the IC_{50} for the anticancer drug cisplatin is 0.7 μM) (Teeyapant *et al.*, 1993). It acts as a ligand dependent protein tyrosine kinase inhibitor of epidermal growth factor receptors (Kreuter *et al.*, 1990). It is thought that the cytotoxicity may in part be due to the fact that aeroplysinin is capable of generating free radicals under physiological conditions (Koulman *et al.*, 1996) The synthesis of aeroplysinin was achieved by Andersen and Faulkner (1975) in a non-stereospecific fashion. Recent studies have shown that sponge grafts can produce aeroplysinin *in vitro*, and under optimal conditions can produce 13 mg of aeroplysinin per 100g of sponge (Kreuter *et al.*, 1992).

6.3 GIROLLINE

Girroline (**2**), also known as girodazole, was isolated from *Pseudaxinyssa cantharella* and *Axinella* sp. (sponges of order Axinellida and Family Axinellidae) from New Caledonia (Ahond *et al.*, 1988; Chiaroni *et al.*, 1991). Girolline contains a chlorine atom, a feature that is rare in terrestrial secondary metabolites. It is active *in vivo* against several murine grafted tumours: leukemia (P388, L1210), mammary adenocarcinoma (MA 16/C) and histiocytosarcoma (M5076) (Lavelle *et al.*, 1991). Its mode of action is as inhibitor of protein synthesis during the elongation/termination steps (Colson *et al.*, 1992). The stereoselective synthesis of girolline has been accomplished (Ahond *et al.*, 1992), and it went into phase-I clinical trials in France in 1991 as an experimental drug under the sponsorship of EORTC and Rhône-Poulenc-Rorer, since toxicological studies on animals did not reveal any major toxic effect that would preclude administration in patients (Lavelle *et al.*, 1991). Since then, trials have been suspended due to blood pressure complications.

6.4 HALICHONDRIN-B

Halichondrin-B (**3**) is a complex polyether macrolide which was first isolated in a $2 \times 10^{-6}\%$ yield from *Halichondria okadai* (sponge of order Halichondrida and family Halichondriidae) collected in Japan (Hirata and Uemura, 1986). Previous studies on *Halichondria okadai* yielded a cytotoxic compound, okadaic acid, a protein phosphatase inhibitor. Halichondrin-B was discovered later in an unrelated sponge *Lissodendoryx* sp. (order Poecilosclerida and family Myxillidae) from New Zealand together with related metabolites (Litaudon *et al.*, 1994). The original publication by Hirata and Uemura showed that halichondrin-B had an IC_{50} of 0.093 ng/ml (84 pM) against B16 melanoma cells. Their *in vivo* studies using B16 melanoma and P388 leukemia gave incredible survival time increases (test/control), 244% and 323%, respectively. It was found that halichondrin-B had an IC_{50} value for L1210 murine

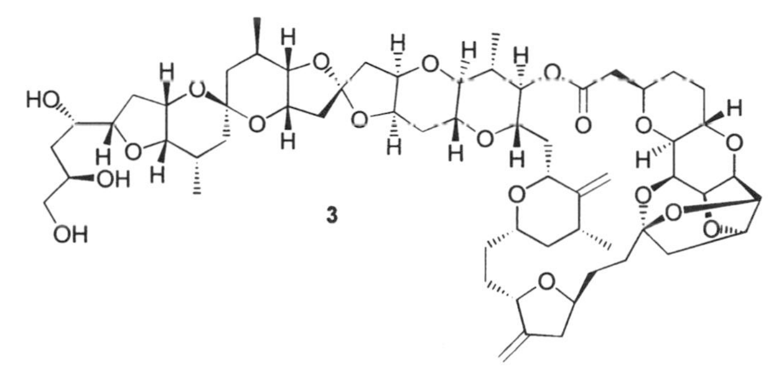

Structure 6.3

leukemia cells of 0.3 nM. The leukemic cells were arrested in mitosis at toxic concentrations, and halichondrin-B inhibited the polymerization of purified tubulin, as well as inhibiting microtubule assembly dependent on microtubule-associated proteins. Also it was found to be a non-competitive inhibitor of vinblastine binding to tubulin, and therefore was compared with drugs with similar modes of action, vinblastine, maytansine, dolastatin-10 (see below), phomopsin-A and rhizoxin (Bai *et al.*, 1991). By analogy with maytansine, which is a competitive inhibitor of vinblastine binding, halichondrin-B has no effect on colchicine binding (Luduena *et al.*, 1993). In addition, at the NCI's 60-cell line screen, halichondrin-B showed activity in the leukemia, non-small-cell lung, colon and ovarian panels. Clinical trials have been hampered by the fact that there was no ready supply of the compound for testing. Synthesis, though possible, was not an option, as this requires many steps and the yield is very low (Aicher *et al.*, 1992). Alternative strategies for synthesis are being worked on by various groups, although these will never provide a viable supply of halichondrin-B (Burke, Zhang and Buchanan, 1995; Horita, Hachiya and Ogihara, 1996). Current supplies for preclinical trials, conducted at the NCI, are being provided by Murray Munro's group at the University of Canterbury, New Zealand in conjunction with Chris Battershill and coworkers at New Zealand's National Institute of Water and Atmospheric Research (NIWA). Initially a survey was conducted and it was found that there were only about 300 tonnes of *Lissodendoryx* sp. in existence, not enough to provide a sustainable supply of halichondrin-B. The group at NIWA are currently attempting to perform aquaculture of *Lissodendoryx* sp. although this work has not yet been published (Pain 1996). It is hoped that this approach will yield a sustainable supply of halichondrin-B. Another approach being taken by Shirley Pomponi and colleagues at Harbor Branch Oceanographic Institution (HBOI) is sponge cell tissue culture of *Lissodendoryx* sp. This work started in early 1996, but has not yet reached the stage of halichondrin-B production *in vitro*, though it is hoped that this will be achieved in the not too distant future (Pomponi, 1997).

6.5 JASPLAKINOLIDE

Jasplakinolide (**4**) was isolated by two Californian groups from the sponge *Jaspis johnstoni* (sponge of order Astrophorida and family Coppatidae) from Fiji (Crews, Manes and Boehler, 1986; Zabriskie *et al.*, 1986). It was isolated later from a Papua New Guinean sponge *Auletta* cf. *constricta* (order Axinellida and family Axinellidae) (Crews *et al.*, 1994b). Jasplakinolide is a depsipeptide structure containing three amino acids, one of which is a brominated tryptophan. The initial papers reported that it had anthelmitic, insecticidal and antifungal properties. Its structure suggests that it should be an ideal complexing agent for metal ions, and it was found to be a lithium complexing agent (Inman, Crews and McDowell, 1989). The asymmetric total synthesis of jasplakinolide was reported in 1991, with an overall yield of 6.6% (Chu, Negrete and Konopelski, 1991) When evaluated in the NCI's 60-cell line screen it was found to have a unique profile. Jasplakinolide was particularly potent *in vitro* against PC3 prostate carcinoma cells with a GI_{50} of 34 nM. It also showed activity in the melanoma (GI_{50} 36 nM), CNS (GI_{50} 37 nM), renal (GI_{50} 52 nM) and lung cancer (GI_{50} 97 nM) cell line panels. It is solid tumour selective with much lower activities being reported than for leukemia and lymphoma. These results made it an attractive candidate for preclinical trials and jasplakinolide was advanced for *in vivo* studies for prostate cancer (PC3) and melanoma (SK-MEL28). A 45% weight reduction of a mouse xenograft PC3 tumour was achieved in 13 days. The mode of activity of jasplakinolide is as actin filament stabilizing agent and it induces the overpolymerization of actin (Senderowicz *et al.*, 1995). A fluorescently labelled jasplakinolide is being marketed by

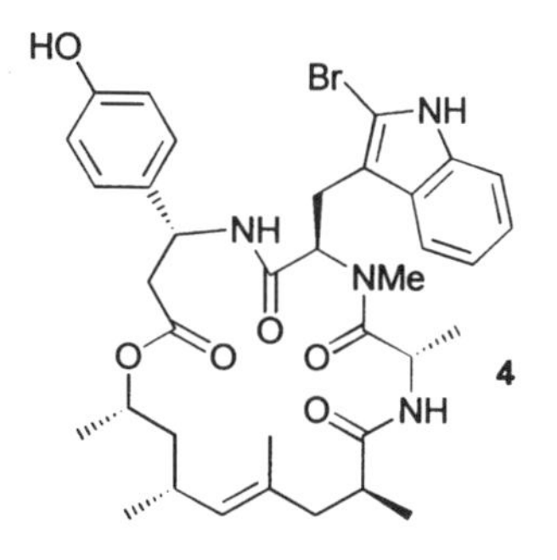

Structure 6.4

Molecular Probes Inc. as a cell permeant actin polymerization inducer for the study of actin polymerization in live cells.

6.6 DIDEMNIN-B

Didemnin-B (**5**) is a cytotoxic and antiviral cyclic depsipeptide isolated from *Trididemnum solidum* (chordate (ascidian) of order Enterogona and family Didemnidae) from the Caribbean (Rinehart *et al.*, 1981a,b). Its total synthesis was not achieved for another 8 years (Hamada *et al.*, 1989) In the initial publications it was found to be active against murine leukemia L1210 *in vitro* (ID_{50} 1.1 ng/ml, 1.1 nM) and P388 murine leukemia and B16 melanoma *in vivo* (survival time increase (test/control) of 199% and 160%, respectively). Also it has activity against B16 melanoma, P388 leukemia and M5076 sarcoma *in vivo* (Stewart *et al.*, 1991) It induces human HL-60 cells to undergo apoptosis to 100% within 140 minutes, with typical apoptotic morphology (Grubb, Wolvetang and Lawen, 1995). Didemnin-B was the first marine natural product to enter clinical trials, and entered Phase II in the mid 1980's (Chun, Wolvetang and Lawen, 1986). It was tested in patients with colorectal, renal, non-small-cell lung, cervical, ovarian, breast and prostate cancers. Problems emerged with several of the trials, because the neuromuscular toxicity of didemnin-B appeared to be dose limiting (Shin *et al.*, 1991) In the case of colorectal cancer, the investigators concluded that didemnin-B did "not compare favourably with results of treatments using other single agents or combinations that are currently available for the treatment of advanced colorectal cancer" (Jones *et al.*, 1992) These side effects caused the trials to be suspended. Rinehart's group have attempted to modify the therapeutic index of the didemnins by carrying out a structure–activity relationship study using natural, synthetic and semi-synthetic didemnin analogues, and the most promising compound appears to be dehydrodidemnin-B (Sakai *et al.*, 1996).

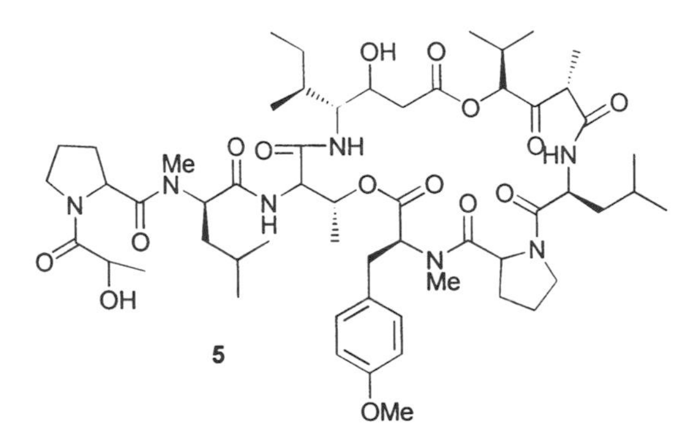

Structure 6.5

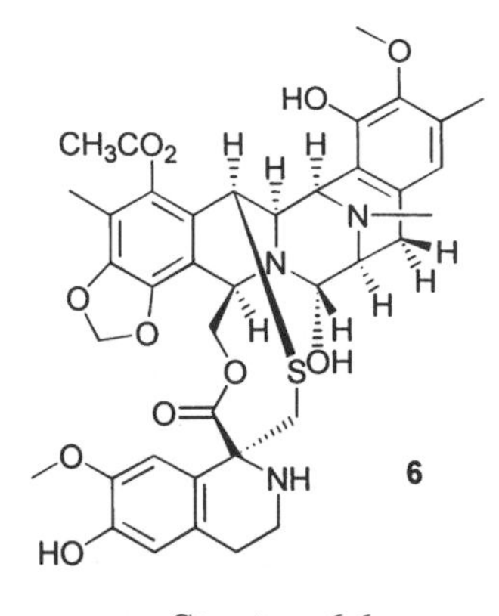

Structure 6.6

6.7 ECTEINASCIDIN 743

Ecteinascidin 743 (**6**) was isolated originally from the tunicate *Ecteinascidia turbinata* (chordate (ascidian) of order Enterogona and family Perophoridae) with a yield of $10^{-4}\%$ from the Caribbean (Rinehart *et al.*, 1990). The complex alkaloid structure has been synthesised only recently, using a highly convergent approach in a 3–4% total yield (Corey, Gin and Kania, 1996). The initial isolation paper reported ecteinascidin as having antitumour properties against various cell lines, L1210 *in vitro* (IC_{50} 0.5 ng/ml, 0.7 nM) and P388 *in vivo* (survival time increase (test/control) of 167%). It was found that ecteinascidin-743 acted as an antimitotic agent, not by binding to tubulin, but by disorganizing the microtubule network in some fashion (Garciarocha, Garciagravalos and Avila, 1996). It was found later that it is a DNA minor groove guanine specific alkylating agent (Pommier *et al.*, 1996) The compound has been licenced by the Spanish company PharmaMar, and is entering Phase I clinical trials in France and Phase I clinical trials in the UK at the Imperial Cancer Research Fund research unit at the Western General Hospital in Edinburgh. Trials are being conducted against small cell lung cancer, melanoma and breast cancer (Smyth, 1997). A current dosage regime suggests three 0.5 mg doses of ecteinascidin-743 per patient. It is estimated that 5000 kg of the organism is needed to provide the 5 g of ecteinascidin-743 needed for the completion of worldwide trials. Current supplies of ecteinascidin-743 are being provided by sustainable collections of *Ecteinascidia turbinata*. The organism is collected from its mangrove root habitat whilst leaving the stolon (root) behind, thus allowing it to regrow, and reproduce sexually (Pain 1996).

6.8 DOLASTATIN-10

Dolastatin-10 (**7**) is a peptide consisting of five amino acid residues, four of which are modified, isolated from *Dolabella auricularia* (mollusc (sea hare) of

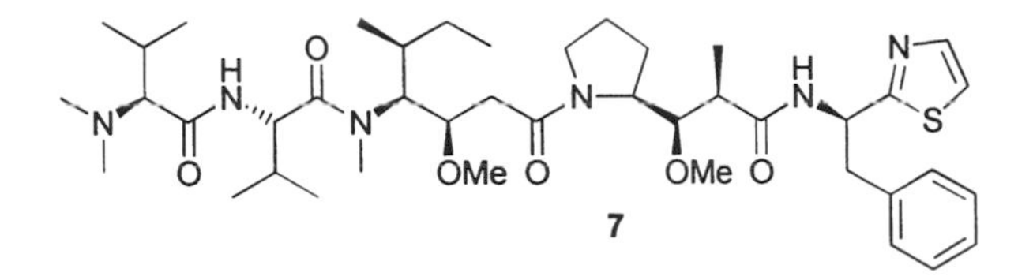

Structure 6.7

the order Anaspidea and family Aplysiidae) collected in Mauritius (Pettit *et al.*, 1987). The initial report indicated that dolastatin-10 was the most active of the dolastatins, achieving a 138% life extension using human B16 melanoma xenografts in mice. In the NCI 60-cell line screen it shows remarkable activity in the lung, colon and CNS cancer panels. Dolastatin-10 inhibited murine and human bone marrow cell colony formation with an ID_{50} of 0.1–1 pg/ml, with complete inhibition occurring at 10–100 pg/ml (Jacobsen *et al.*, 1991). It is found to be more potent than vinblastine or taxol with an IC_{50} of 0.23 nM against human ovarian and colon cancer cell lines (Aherne *et al.*, 1996). Its biological activity made it an extremely interesting potential anticancer drug, but the isolated yield of 10^{-6}–10^{-7}% made it unfeasible to collect the sea hare in bulk to obtain enough dolastatin-10 for preclinical and clinical trials. The isolation procedure to purify enough dolastatin-10 for structural elucidation required the collection of 1600 kg *Dolabella auricularia* (Pettit *et al.*, 1993). This made the total synthesis a high priority, and it was achieved in a stereoselective fashion relatively rapidly (Hamada, Hayashi and Shioiri, 1991; Shioiri, Hayashi and Hamada, 1993; Roux *et al.*, 1994). These syntheses also allowed a number of analogues to be synthesized and tested for activity (Pettit *et al.*, 1995). Synthetic dolastatin-10 will be used for the Phase-I clinical trials, which will be starting shortly at the NCI. Dolastatin-10 is an antimitotic (tubulin polymerization inhibitor) and non-competitively inhibits the binding of vinca alkaloids to tubulin, and most resembles phomopsin-A in its activity (Bai, Pettit and Hamel, 1990; Luduena *et al.*, 1992).

6.9 BRYOSTATIN-1

Bryostatin-1 (**8**) was isolated first from the seamat (bryozoan) *Bugula neritina* (order Cheilostomata and family Bugulidae) from the eastern Pacific Ocean (California) (Pettit *et al.*, 1982). The NMR data published initially were revised (Schaufelberger, Chmurny and Koleck, 1991). In the NCI's 60-cell line screen bryostatin-1 showed activity in the lung, colon, CNS, melanoma, ovarian and renal cancer panels. It is unusual in its activity as instead of damaging bone marrow progenitor cells like many anticancer drugs, bryostatin-1 is a powerful stimulator of these cells (May *et al.*, 1987). It has potential to be used in combination chemotherapy to counteract the myelosuppression associated

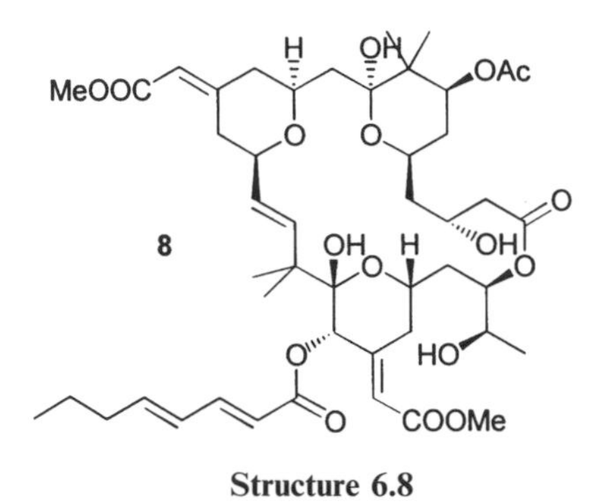

Structure 6.8

with some chemotherapy agents. Bryostatin-1 has been noted also for its immunoenhancing properties, and it activates and induces the proliferation of T and B cells (McGown, 1995). Bryostatin-1 is a powerful activator of the signal transduction enzyme protein kinase C (PKC), which has been implicated in the abnormal cell signalling observed in cancerous cells (Szallasi *et al.*, 1994). It is these multiple activities displayed by bryostatin-1 that make it such an exciting candidate for an anticancer drug. Phase I clinical trials have been completed in the UK under the sponsorship of the Cancer Research Campaign at the Christie Hospital in Manchester (Philip *et al.*, 1993; Jayson *et al.*, 1995). The Phase I trials showed that myalgia was the dose limiting toxicity, but positive responses were noted against metastatic melanoma, non-Hodgkin's lymphoma and ovarian cancer, and as a result of this Phase II studies are planned. Similar results were obtained in the NCI sponsored Phase-I clinical trials in the USA, and solicitations for Phase II trials were made by the NCI for renal carcinoma, melanoma, chronic lymphocytic leukemia, non-Hodgkin's lymphoma, myeloma, ovarian, uterine and cervical cancers. The supplies of bryostatin-1 will be met by aquaculture, as a viable total synthesis is not yet complete, though studies are at an advanced stage (Hale *et al.*, 1995; Debrabander and Vandewalle, 1996). Under the sponsorship of the NCI, Calbiomarine Technologies Inc. in California is hoping to produce reasonable supplies of bryostatin-1 by aquaculture (Olson, 1996). Normal yields of bryostatin-1 from natural populations of *Bugula neritina* are in the region of 10^{-6}% wet weight, and projected demand for bryostatin-1, if approved for drug use, is of the order of 100–500 g/year, whereas it is estimated that sustainable collection of wild populations will yield only 30–40 g/year. The aquaculture system at Calbiomarine involves leaving culture plates settled with *B. neritina* larvae in a land based tank for 3–4 weeks after which the plates are left in the open ocean for a period of 5–6 months. It is estimated that an aquaculture system capable of producing 100 g of bryostatin-1 per year would require an initial expenditure of US$ 4 M and US$ 2 M per year to operate. The current yields of *B. neritina* are 2.0–3.8 kg wet weight per square meter of culture plate, with a yield of bryostatin-1 of about 7.0 μg per g dry weight. It is possible to increase the amount of bryostatin-1 harvested by synthetically converting bryostatin-2 to bryostatin-1 (Pettit *et al.*, 1991). Bryostatin-1 is perhaps the most advanced in clinical trials of all the compounds described in this chapter, and it is hoped that it will be approved as an anticancer drug.

6.10 HALOMON

Halomon (**9**), a highly halogenated monoterpene, was isolated from a red alga *Portieria hornemanni* (order Gigartinales and family Rhizophyllidacea) from the Phillipines (Fuller *et al.*, 1992). The NCI's 60-cell line screen showed that it was active mainly against brain, renal and colon tumour cell lines, with a mean panel GI_{50} of 0.67 mM. Comparison with the database of known anticancer agents showed halomon to have a unique activity profile, although its mode of action has not yet been elucidated. Further preclinical work was hampered severely due to the inability to procure a collection of the organism which consistently produced halomon (Fuller *et al.*, 1994), and the fact that an asymmetric synthesis has not yet been achieved. Bioavailability studies using halomon showed that it was widely distributed to all tissues, but that it was concentrated and persisted in the fatty tissues (Egorin *et al.*, 1996). Another observation was that halomon persisted at the site of injection due to its low aqueous solubility. The problems of supply and bioavailability have led to further clinical trials being delayed.

6.11 FUTURE DIRECTIONS

The above survey has highlighted the unparalleled potential of marine natural products as anticancer therapeutics. In many of these cases the extreme structural novelty coupled with an unusual mode of biological activity make these compounds attractive as anticancer therapeutics. The availability of a sustainable supply has been addressed in some cases, with synthesis (e.g., dolastatin-10) and aquaculture (e.g., bryostatin-1) expected to meet demands for clinical usage. The application of cell tissue culture is expected to improve in the near

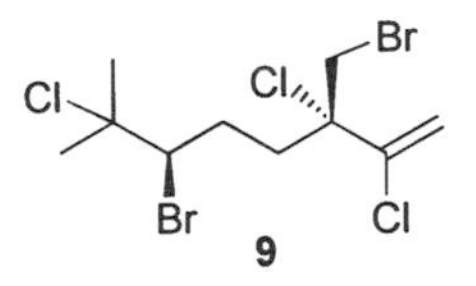

Structure 6.9

future, and it is hoped this will lead to the production of some of the compounds described above (e.g. halichondrin-B). However, there are examples in which a sustainable source has not been pinpointed. In many cases, it is suspected that the compound responsible for bioactivity is not produced by the macroorganism, but by a symbiont microorganism (Kobayashi and Ishibashi, 1993). This is a distinct possibility for the highly bioactive manzamine alkaloids, which are found in sponges from very different taxonomic divisions (Crews *et al.*, 1994a). It has been suggested that all these sponges contain related symbionts, and that these are responsible for the production of the complex metabolites. The evidence for symbiotic production of bioactive secondary metabolites is strong in some cases (Bewley, Holland and Faulkner, 1996), and this has led to investigations of natural products from marine microorganisms (Fenical and Jensen, 1993; Fenical, 1993; Jensen and Fenical, 1994). In this case the microorganism can be cultured to provide a sustainable supply of the compound of interest for clinical usage. Sponges often are infested with filamentous cyanobacteria, and investigations of the filamentous cyanobacterium *Lyngbya majuscula* (order Nostocales and family Oscillatoriaceae) collected in Curaçao gave the antimitotic and antiproliferative compound curacin-A (**10**) (Gerwick *et al.*, 1994). It is active *in vitro* against L1210 leukemia (IC_{50} 9 nM) and CA46 Burkit lymphoma (IC_{50} 0.2 mM). Curacin-A is an antimitotic and inhibits the polymerization of purified tubulin, induced by either glutamate or microtubule associated protein dependent microtubule assembly. Curacin-A currently is under pre-clinical investigation, and supplies are being provided by laboratory cultures of *Lyngbya majuscula.*

The above review will have alerted the reader to the fact that natural products isolated from marine organisms have, and will continue to provide, novel pharmacophores to be used directly as pharmaceuticals, or to be modified by medicinal and combinatorial chemistry methods. In my opinion, combinatorial methods are best applied to modify lead compounds produced by nature to achieve a better therapeutic index. Secondary metabolites therefore still occupy a very important position in the generation of novel lead compounds, and the correct balance must be struck between the use of natural products and synthetic chemistry in order to afford optimum development of therapeutics for the treatment of cancer and infectious diseases. It is found that secondary metabolites produced by terrestrial organisms tend to be restricted to a

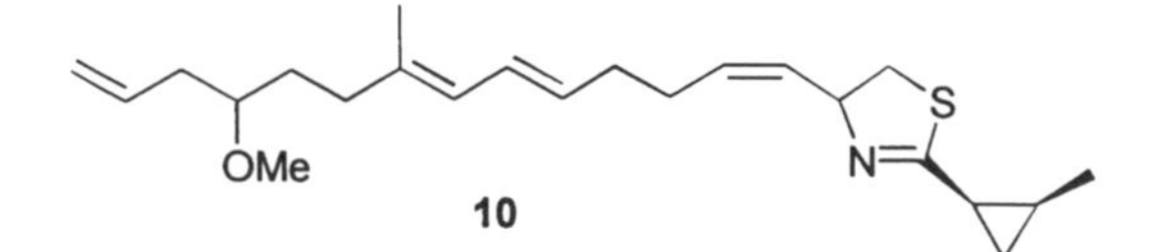

Structure 6.10

known series of structural types, and that many compounds isolated from these organisms have been reported previously in the literature. Only a small percentage of all marine organisms have been investigated for their potential to produce novel pharmacophores. There are many as yet undiscovered and inaccessible species which will have evolved under unusual selectional pressures that have the potential to yield novel structural types with unprecedented modes of biological activity.

REFERENCES

Aherne, G.W., Hardcastle, A., Valenti, M., Bryant, A., Rogers, P. and Pettit, G.W. (1996). Antitumor evaluation of dolastatin-10 and dolastatin-15 and their measurement in plasma by radioimmunoassay, *Cancer Chemotherapy and Pharmacology*, **38**, 225–32.

Ahond, A., Almourabit, A., Bedoyazurita, M., Heng, R., Braga, R.M. and Poupat, C. (1992) Stereoselective synthesis of girolline, *Tetrahedron*, **48**, 4327–46.

Ahond, A., Zurita, M.B., Colin, M., Fizames, C., Laboute, P. and Lavelle, F. (1988) Girolline, a new antitumoral compound extracted from the sponge *Pseudaxinyssa cantharella*, *Comptes Rendus de l'Academie des Sciences* II, **307**, 145–8

Aicher, T.D., Buszek, K.R., Fang, F.G., Forsyth, C.J., Jung, S.H. and Kishi, Y. (1992) Total synthesis of halichondrin-B and norhalichondrin-B, *Journal of the American Chemical Society*, **114**, 3162–4.

Andersen, R.J., and Faulkner, D.J. (1975) Synthesis of aeroplysinin-1 and related compounds, *Journal of the American Chemical Society*, **97**, 936–7.

Aszalos, A. (ed.) (1981) *Antitumor Compounds of Natural Origin: Chemistry and Biochemistry*. CRC Press, Boca Raton.

Bai, R., Paull, K.D., Herald, C.L., Malspeis, L. Pettit, G.R. and Hamel, E. (1991) Halichondrin-B and homohalichondrin-B, marine natural-products binding in the vinca domain of tubulin, *Journal of Biological Chemistry*, **266**, 15882–9.

Bai, R., Pettit, G.R. and Hamel, E. (1990) Binding of dolastatin-10 to tubulin at a distinct site for peptide antimitotic agents near the exchangeable nucleotide and vinca alkaloid sites, *Journal of Biological Chemistry*, **265**, 17141–9.

Bewley, C.A., Holland, N.D. and Faulkner, D.J. (1996) Two classes of metabolites from *Theonella swinhoei* are localised in distinct populations of bacterial symbionts, *Experientia*, **52**, 716–22.

Blunt, J.W. and Munro, M.H.G. (1996) MarinLit vpc1.2 (Windows), Canterbury, New Zealand.

Bongiorni, L. and Pietra, F. (1996) Marine natural products for industrial applications, *Chemistry and Industry*, 15 Jan, 54–8.

Buckingham, J. (ed.) (1997) *Dictionary of Natural Products*, Vol. 6.1, on CD-ROM, Chapman & Hall, London.

Burke, S. D., Zhang, G. and Buchanan, J.L. (1995) Enantioselective synthesis of a halichondrin-B C(20)-C(36) precursor, *Tetrahedron Letters*, **36**, 7023–6.

Chiaroni, A., Riche, C. Ahond, A., Poupat, C., Pusset, M. and Potier, P. (1991) Crystal-structure and absolute configuration of girolline, *Comptes Rendus de l' Academie des Sciences* II, **312**, 49–53.

Chu, K. S., Negrete, G.R. and Konopelski, J.P. (1991) Asymmetric total synthesis of (+)-jasplakinolide, *Journal of Organic Chemistry*, **56**, 5196–202.

Chun, H. G., Davies, B., Hoth, D., Suffness, M., Plowman, J. and Flora, K. (1986) Didemnin-B – the 1st marine compound entering clinical-trials as an antineoplastic agent, *Investigational New Drugs*, **4**, 279–84.

Cimino, G., De Rosa, S., De Stefano, S., Self, R. and Sodano, G. (1983) The bromo compounds of the true sponge *Verongia aerophoba*, *Tetrahedron Letters*, **24**, 3029–32.

Colin, P.L. and Arneson, C. (1995) *Tropical Pacific Invertebrates*, Coral Reef Press, Beverly Hills, CA.

Colson, G., Rabault, B., Lavelle, F. and Zerial, A. (1992) Mode of action of the antitumor compound girodazole, *Biochemical Pharmacology*, **43**, 1717–23.

Corey, E. J., Gin, D.Y. and Kania, R.S. (1996) Enantioselective total synthesis of ecteinascidin-743, *Journal of the American Chemical Society*, **118**, 9202–3.

Cragg, G.M., Newman, D.J. and Snader, K.M. (1997) Natural products in drug discovery and development, *Journal of Natural Products*, **60**, 52–60.

Crews, P., Cheng, X.C., Adamczeski, M., Rodriguez, J., Jaspars, M., Schmitz, F.J., Traeger, S.C. and Pordesimo, E.O. (1994a) 1,2,3,4-Tetrahydro-8-hydroxymanzamines, alkaloids from 2 different haplosclerid sponges, *Tetrahedron*, **50**, 13567–74.

Crews, P., Farias, J.J., Emrich, R. and Keifer, P. (1994b) Milnamide A, an unusual cytotoxic tripeptide from the marine sponge *Auletta* cf *constricta*, *Journal of Organic Chemistry*, **59**, 2932–4.

Crews, P., Manes, L.V. and Boehler, M. (1986) Jasplakinolide, a cyclodepsipeptide from the marine sponge, *Jaspis* sp. *Tetrahedron Letters*, **27**, 2797–800.

Crews, P., Rodriguez, J. and Jaspars, M. (1998) *Organic Structure Analysis*, Oxford University Press, New York, in press.

Debrabander, J. and Vandewalle, M. (1996) Towards the asymmetric synthesis of bryostatin-1, *Pure and Applied Chemistry*, **68**, 715–18.

Ebel, R., Brenzinger, M., Kunze, A., Gross, H.J. and Proksch, P. (1997) Wound activation of protoxins in marine sponge *Aplysina aerophoba*, *Journal of Chemical Ecology*, **23**, 1451–62.

Egorin, M. J., Sentz, D.L., Rosen, D.M., Ballesteros, M.F., Kearns, C.M. and Callery, P.S. (1996) Plasma pharmacokinetics, bioavailability, and tissue distribution in CD2F1 mice of halomon, *Cancer Chemotherapy and Pharmacology*, **39**, 51–60.

Fattorusso, E., Minale, L. and Sodano, G. (1970) Aeroplysinin-1; a bromo compound from the sponge *Aplysina aerophoba*, *Journal of the Chemical Society; Chemical Communications*, 751–3.

Fattorusso, E., Minale, L. and Sodano, G. (1972) Aeroplysinin-1, an antibacterial bromo compound from the sponge *Verongia aerophoba*, *Journal of the Chemical Society; Perkin Transactions* I, 16–18.

Faulkner, D.J. (1993) Academic chemistry and the discovery of bioactive marine natural products, in *Marine Biotechnology* (eds D.H. Attaway and O.R. Zaborsky), Plenum Presss, New York.

Fenical, W. (1993) Chemical studies of marine bacteria: developing a new resource, *Chemical Reviews*, **93**, 1673–83.

Fenical, W., and Jensen, P.R. (1993) Marine microorganisms: a new biomedical resource, in *Marine Biotechnology* (eds D.H. Attaway and O.R. Zaborsky), Plenum Press, New York.

Fuller, R.W., Cardellina, J.H., Jurek, J., Scheuer, P.J. and Alvaradolindner, B. (1994) Isolation and structure-activity features of halomon-related antitumor monoterpenes from the red alga *Portieria hornemannii*, *Journal of Medicinal Chemistry*, **37**, 4407–11.

Fuller, R.W., Cardellina, J.H., Kato, Y., Brinen, S., Clardy, J., Snader, K.M. and Boyd, M.R. (1992) A pentahalogenated monoterpene from the red alga *Portieria hornemannii* produces a novel cytotoxicity profile against a diverse panel of human tumor cell lines, *Journal of Medicinal Chemistry*, **35**, 3007–11.

Garciarocha, M., Garciagravalos, M. and Avila, J. (1996) Characterization of antimitotic products from marine organisms that disorganise the microtubule network ecteinascidin-743, *British Journal of Cancer*, **73**, 875–83.

Garson, M.J. (1994) The biosynthesis of sponge secondary metabolites, why it is important, in *Sponges in Time and Space* (eds R.W.M. van Soest, T.M.G. van Kempen and J.C. Braekman), Balkema, Rotterdam.

Gerwick, W.H., Proteau, P.J., Nagle, D.G., Hamel, E., Blokhin, A. and Slate, D.L. (1994) Structure of Curacin-A, a novel antimitotic, antiproliferative, and brine shrimp toxic natural product from the marine cyanobacterium *Lyngbya majuscula*, *Journal of Organic Chemistry*, **59** 1243–5.

Grubb, D.R., Wolvetang, E.J. and Lawen, A. (1995) Didemnin-B induces cell-death by apoptosis – the fastest induction of apoptosis ever described, *Biochemical and Biophysical Research Communications*, **215**, 1130–6.

Hale, K.J., Lennon, J.A., Manaviazar, S., Javaid, M.H. and Hobbs, C.J. (1995) Asymmetric-synthesis of the C(17)-C(27) segment of the antineoplastic macrolide bryostatin-1, *Tetrahedron Letters*, **36**, 1359–62.

Hamada, Y., Hayashi, K. and Shioiri, T. (1991) Efficient stereoselective synthesis of dolastatin-10, an antineoplastic peptide from a sea hare, *Tetrahedron Letters*, **32**, 931–4.

Hamada, Y., Kondo, Y., Shibata, M. and Shioiri, T. (1989) Efficient total synthesis of didemnins A and B, *Journal of the American Chemical Society*, **111**, 669–73.

Hirata, Y., and Uemura, D. (1986) Halichondrins-antitumor polyether macrolides from a marine sponge, *Pure and Applied Chemistry*, **58**, 701–10.

Horita, K., Hachiya, S. and Ogihara, K. (1996) Synthetic studies of halichondrin-B, an antitumor polyether macrolide, *Heterocycles*, **42**, 99–104.

Inman, W., Crews, P. and McDowell, R. (1989) Novel marine sponge derived amino-acids. 9. Lithium complexation of jasplakinolide, *Journal of Organic Chemistry*, **54**, 2523–6.

Ireland, C.M., Copp, B.R., Foster, M.P., McDonald, L.A., Radisky, D.C. and Swersey, J.C. (1993) Biomedical potential of marine natural products, in *Marine Biotechnology* (eds D.H. Attaway and O.R. Zaborsky), Plenum Press, New York.

Jacobsen, S.E.W., Ruscetti, F.W., Longo, D.L. and Keller, J.R. (1991) Antineoplastic dolastatins – potent inhibitors of hematopoietic progenitor cells, *Journal of the National Cancer Institute*, **83**, 1672–7.

Jaspars, M., Rali, T. Laney, M. Schatzman, R.C., Diaz, M.C., Schmitz, F.J., Pordesimo, E.O. and Crews, P. (1994). The search for inosine 5′-phosphate dehydrogenase (IMPDH) inhibitors from marine sponges – evaluation of the bastadin alkaloids, *Tetrahedron*, **50**, 7367–74.

Jayson, G.C., Crowther, D., Prendiville, J., McGown, A.T., Scheid, C., Stern, P., Young, R., Brenchley, P., Chang, J., Owens, S. and Pettit, G.R. (1995) A phase-I trial of bryostatin-1 in patients with advanced malignancy using a 24 hour intravenous infusion, *British Journal of Cancer*, **72**, 461–8.

Jensen, P.R. and Fenical, W. (1994) Strategies for the discovery of secondary metabolites from marine bacteria: ecological perspectives, *Annual Reviews of Microbiology*, **48**, 559–84.

Jones, D.V., Ajani, J.A., Blackburn, R., Daugherty, K., Levin, B. and Patt, Y.Z. (1992) Phase-II study of didemnin-B in advanced colorectal-cancer, *Investigational New Drugs*, **10**, 211–13.

Kobayashi, J. and Ishibashi, M. (1993) Bioactive metabolites of symbiotic marine microorganisms, *Chemical Reviews*, **93**, 1753–69.

Koulman, A., Proksch, P., Ebel, R., Beekman, A.C., Vanuden, W. and Konings, A.W.T. (1996) Cytotoxicity and mode of action of aeroplysinin-1 and a related dienone from the sponge *Aplysina aerophoba*, *Journal of Natural Products*, **59**, 591–4.

Kreuter, M.H., Bernd, A., Holzmann, H., Mullerklieser, W. and Maidhof, A. (1989) Cytostatic activity of aeroplysinin-1 against lymphoma and epithelioma cells, *Zeitschrift fur Naturforschung C*, **44**, 680–8.

Kreuter, M.H., Leake, R.E., Rinaldi, F., Mullerklieser, W., Maidhof, A., Muller, W.E.G. and Schroder, H.C. (1990) Inhibition of intrinsic protein tyrosine kinase-activity of EGF-receptor kinase complex from human breast-cancer cells by the marine sponge metabolite aeroplysinin-1, *Comparative Biochemistry and Physiology* B, *Comparative Biochemistry*, **97**, 151–8.

Kreuter, M.H., Robitzki, A., Chang, S., Steffen, R., Michaelis, M. and Kljajic, Z. (1992) Production of the cytostatic agent aeroplysinin by the sponge *Verongia aerophoba* in *in vitro* culture, *Comparative Biochemistry and Physiology*, C, *Pharmacology Toxicology and Endocrinology*, **101**, 183–7.

Lavelle, R.A., Zerial, A., Fizames, C., Rabault, B. and Curaudeau, A. (1991) Antitumour activity and mechanism of action of the marine compound girodazole, *Investigational New Drugs*, **9**, 233–44.

Litaudon, M., Hart, J.B., Blunt, J.W., Lake, R.J. and Munro, M.H.G. (1994) Isohomohalichondrin-B, a new antitumor polyether macrolide from the New Zealand deep water sponge *Lissodendoryx* sp., *Tetrahedron Letters*, **35**, 9435–8.

Luduena, R.F., Roach, M.C., Prasad, V. and Pettit, G.R. (1992) Interaction of dolastatin-10 with bovine brain tubulin, *Biochemical Pharmacology*, **43**, 539–43.

Luduena, R.F., Roach, M.C., Prasad, V. and Pettit, G.R. (1993) Interaction of halichondrin-B and homohalichondrin-B with bovine brain tubulin, *Biochemical Pharmacology*, **45**, 421–7.

May, W.S., Sharkis, S.J., Esa, A.H., Gebbia, V., Kraft, A.S., Pettit, G.R. and Sensenbrenner, L.L. (1987) Antineoplastic bryostatins are multipotential stimulators of human hematopoietic progenitor cells, *Proceedings of the National Academy of Sciences of the USA*, **84**, 8483.

McGown, A. (1995) Bryostatin, in *Christie Hospital/Paterson Institute Annual Research Report*, pp. 14–15.

Munro, M.H.G., Blunt, J.W., Lake, R.J., Litaudon, M., Battershill, C.N. and Page, M.J. (1994) From seabed to sickbed, what are the prospects?, in *Sponges in Time and Space* (eds R.W.M. van Soest, T.M.G. van Kempen and J.C. Braekman), Balkema, Rotterdam.

Olson, S.G. (1996) Curing cancer through aquaculture, *Sea Technology*, August, 89–94.

Pain, S. (1996) Hostages of the deep, *New Scientist*, 14 September, 38–42.

Pettit, G.R., Herald, C.L., Doubek, D.L., Herald, D.L., Arnold, E. and Clardy, J. (1982) Anti-neoplastic agents 86: isolation and structure of bryostatin-1, *Journal of the American Chemical Society*, **104**, 6846–8.

Pettit, G.R., Kamano, Y., Herald, C.L., Fujii, Y., Kizu, H., Boyd, M.R. and Boettner, F.E. (1993) Isolation of dolastatins 10-15 from the marine mollusc *Dolabella auricularia*, *Tetrahedron*, **49**, 9151–70.

Pettit, G.R., Kamano, Y., Herald, C.L., Tuinman, A.A., Boettner, F.E. and Kizu, H. (1987) Antineoplastic agents 136: the isolation and structure of a remarkable marine animal antineoplastic constituent-dolastatin-10, *Journal of the American Chemical Society*, **109**, 6883–5.

Pettit, G.R., Sengupta, D., Herald, C.L., Sharkey, N.A. and Blumberg, P.M. (1991) Synthetic conversion of bryostatin-2 to bryostatin-1, *Canadian Journal of Chemistry*, **69**, 856–60.

Pettit, G.R., Srirangam, J.K., Barkoczy, J., Williams, M.D., Durkin, K.P.M. and Boyd, M.R. (1995) Antineoplastic agents 337: synthesis of dolastatin-10 structural modifications, *Anti-Cancer Drug Design*, **10**, 529–44.

Philip, P.A., Rea, D., Thavasu, P., Carmichael, J., Stuart, N.S.A., Rockett, H., Talbot, D.C., Ganesan, T., Pettit, G.R., Balkwill, F. and Harris, A.L. (1993) Phase I study of bryostatin-1: assessment of IL-6 and TNF-a induction *in vivo, Journal of the National Cancer Institute*, **85**, 1812–18.

Pommier, Y., Kohlhagen, G., Bailly, C., Waring, M., Mazumder, A. and Kohn, K.W. (1996) DNA sequence-selective and structure-selective alkylation of guanine N2 in the DNA minor groove by ecteinascidin-743, *Biochemistry*, **35**, 13303–9.

Pomponi, S. (1997) Cell tissue culture of *Lissodendoryx* sp., personal communication.

Rinehart, K.L., Gloer, J.B., Cook, J.C., Miszak, S.A. and Scahill, T.A. (1981a) Structures of the didemnins, antiviral and cytotoxic depsipeptides from a Caribbean tunicate, *Journal of the American Chemical Society*, **103**, 1857–9.

Rinehart, K.L., Gloer, J.B., Hughes, R.G., Renis, H.E., McGovern, J.P., Swynenberg, E.B., Stringfellow, D.A., Kuentzel, S.L. and Li, L.H. (1981b) Didemnins, antiviral and antitumor depsipeptides from Caribbean tunicates, *Science*, **212**, 933–5.

Rinehart, K.L., Holt, T.G., Fregeau, N.L., Stroh, J.G., Keifer, P.A., Sun, F. and Li, L.H. (1990) Ecteinascidin-729, 743, 745, 759A, 759B and 770: potent antitumor agents from the Caribbean tunicate *Ecteinascidia turbinata*, *Journal of Organic Chemistry*, **55**, 4512–15.

Roux, F., Maugras, I., Poncet, J., Niel, G. and Jouin, P. (1994) Synthesis of dolastatin-10 and [R-Doe]-dolastatin-10, *Tetrahedron*, **50**, 5345–60.

Sakai, R., Rinehart, K.L., Kishore, V., Kundu, B., Faircloth, G. and Gloer, J.B. (1996) Structure – activity relationships of the didemnins, *Journal of Medicinal Chemistry*, **39**, 2819–34.

Schaufelberger, D.E., Chmurny, G.N. and Koleck, M.P. (1991) H-1 and C-13 NMR assignments of the antitumor macrolide bryostatin-1, *Magnetic Resonance in Chemistry*, **29**, 366–74.

Schmitz, F.J., Bowden, B.F. and Toth, S.I. (1993) Antitumor and cytotoxic compounds from marine organisms, in *Marine Biotechnology* (eds D.H. Attaway and O.R. Zaborsky), Plenum Press, New York.

Senderowicz, A.M.J., Kaur, G., Sainz, E., Laing, C., Inman, W.D. and Rodriguez, J. (1995) Jasplakinolides – inhibition of the growth of prostate carcinoma-cells *in vitro* with disruption of the actin cytoskeleton, *Journal of the National Cancer Institute*, **87**, 46–51.

Shin, D. M., Holoye, P.Y., Murphy, W.K., Forman, A., Papasozomenos, S.C. and Hong, W.K. (1991) Phase-I/II clinical-trial of didemnin-B in non-small-cell lung-cancer, *Cancer Chemotherapy and Pharmacology*, **29**, 145–9.

Shioiri, T., Hayashi, K. and Hamada, Y. (1993) Stereoselective synthesis of dolastatin 10 and its congeners, *Tetrahedron*, **49**, 1913–24.

Smyth, J.F. (1997) Clinical trials of ecteinascidin-743 at ICRF Edinburgh, personal communication.

Stewart, J.A., Low, J.B., Roberts, J.D. and Blow, A. (1991) A Phase-I clinical-trial of didemnin-B, *Cancer*, **68**, 2550–4.

Stone, M.J. and Williams, D.H. (1992) On the evolution of functional secondary metabolites (natural products), *Molecular Microbiology*, **6**, 29–34.

Szallasi, Z., Denning, M.F., Smith, C.B., Dlugosz, A.A., Yuspa, S.H. and Pettit, G.R. (1994) Bryostatin 1 protects protein kinase-C-δ from down-regulation in mouse keratinocytes in parallel with its inhibition of phorbol ester induced differentiation, *Molecular Pharmacology*, **46**, 840–50.

Teeyapant, R., Woerdenbag, H.J., Kreis, P., Hacker, J., Wray, V. and Witte, L. (1993) Antibiotic and cytotoxic activity of brominated compounds from the marine sponge *Verongia aerophoba, Zeitschrift fur Naturforschung C*, **48**, 939–45.

Zabriskie, T.M., Klocke, J.A., Ireland, C.M., Marcus, A.H., Molinski, T.F., Faulkner, D.J., Xu, C. and Clardy, J.C. (1986) Jaspamide, a modified peptide from a *Jaspis* sponge with insecticidal and antifungal activity, *Journal of the American Chemical Society*, **108**, 3123–4.

7

Drug Discovery by Using Molecular Biology to Reprogramme Microorganisms

Iain S. Hunter

7.1 INTRODUCTION

Streptomyces are soil bacteria which have provided the majority of clinically useful antibiotics, including streptomycin, the tetracyclines, erythromycin and some of the new generation of beta-lactam antibiotics, based on altered penicillins. The advent of target-based screening to detect novel pro-drugs coupled with the number of isolates that can be processed by high-throughput screening has shown that the *Streptomyces* also make a number of metabolites which, although they are poor anti-infective agents (i.e., poor antibiotics), have other biological activities that make them useful as therapeutic agents. For example, some of the leading anticancer drugs are of streptomycete origin. Avermectin, the market-leading endectocide used to eradicate parasites in some domesticated animals, is a streptomycete product. Metabolites with immunosuppressant activity, protease inhibitors and vasodilators have all been discovered in *Streptomyces*. Some examples, which are in clinical use or undergoing clinical trials, are listed in Table 7.1.

Since the 1950's, the *Streptomyces* and their antibiotics have been the infantry in the battle to control infection. But, clinical infections are steadily acquiring resistance to existing antibiotics, and this brings into focus the need to discover new anti-infective agents to combat these antibiotic-resistant strains. Industry is using high-throughput strategies in their discovery programmes to screen novel soil isolates of *Streptomyces* and their close relatives in the Actinomycete family, in the hope of finding the next generation of antibiotic structures. However, a second strategy is to exploit knowledge of

Advances in Drug Discovery Techniques. Edited by Alan L. Harvey
ISBN 0 471 97509 5

Table 7.1 Some examples of streptomycete therapeutic agents, the species which produces them and their biological activity

Product	*Producer micro-organism*	*Biological activity*
Oxytetracycline	*S. rimosus*	anti-infective
Erythromycin	*S. erythreus*	anti-infective
Tetracenomycin	*S. glaucescens*	anticancer
Adriamycin	*S. peucetius*	anticancer
Mithramycin	*S. galilaeus*	anticancer
Avermectin	*S. avermetilis*	antiparasite
Rapamycin	*S. hygroscopicus*	immunosuppressant

the biosynthesis of existing streptomycete metabolites for the rational design of novel chemical structures ('novel chemical entities', as they are commonly known), which might have utility as anti-infective agents or be potential therapeutic agents.

The common theme among the metabolites in Table 7.1 is that they are members of the polyketide family of chemical structures, and although they have quite different biological activities, they share a common chemistry for their biosynthesis. Polyketides are synthesized in the cell by a complex array of enzymes, whose identities and interactions with each other specify the structure that is made. Thus, the genes which encode these enzymes must contain the blueprint for the structure to be made. By understanding this blueprint, the potential to modify it and thus derive a novel structure may be realized.

This chapter uses oxytetracycline as the model to review our current understanding of the genetic programming of biosynthesis of one subclass of polyketides (the 'aromatic' polyketides), and show how, through genetic reprogramming of *Streptomyces*, it is now possible to derive 'novel chemical entities'.

7.2 STREPTOMYCETE GENETICS AND MOLECULAR BIOLOGY

The initial experiments which led to a gene cloning capability for *Streptomyces* began in 1977, and the system is now well developed (Hopwood *et al.*, 1985). Genes can be cloned from one streptomycete species, characterized and then introduced into another. Analysis of antibiotic production genes over the last 15 years has shown that the genes for biosynthesis of a particular antibiotic are always clustered together on the chromosome. As the primary function of antibiotics is to kill bacterial cells, and the streptomycetes are themselves bacteria, the streptomycetes must be resistant to the antibiotics that they produce: otherwise they would commit suicide. Each antibiotic-production gene cluster invariably also contains at least one resistance gene for the antibiotic that it makes. This provides a facile route to isolate these gene clusters for the first time. Resistance genes can be isolated easily by cloning into a host which itself is sensitive to the antibiotic and selecting for acquisition of the resistance phenotype. The biosynthetic genes are found adjacent to the resistance gene. A large number of antibiotic gene clusters have been cloned using this strategy.

Incidentally, the fact that antibiotic-producing bacteria also contain a gene for resistance to that antibiotic could be the Achilles heel for those pharmaceutical companies who wish to derive new antibiotics solely through screening of soil isolates. The streptomycetes are a reservoir of resistance genes for natural antibiotics, which can be transferred on to the bacteria which cause clinical infections. Thus, any new natural antibiotic, of streptomycete origin, introduced into the clinic is at risk of being rendered useless if infective bacteria pick up the resistance gene from the producer in the environment. The use of molecular biology to derive 'non-natural' antibiotics (these 'novel chemical entities') may circumvent this problem.

7.3 THE CHEMISTRY OF POLYKETIDE BIOSYNTHESIS

Polyketides are derived from the sequential addition of 'extender' units to a 'starter' unit. In the case of oxytetracycline (OTC; Figure 7.1) a starter unit containing three carbons is extended by the sequential addition of 8 two-carbon acetate units, to result in the nineteen-carbon skeleton which forms the backbone of the OTC molecule. Note that, as the skelton is built up, each alternate carbon atom has a =O group attached. This keto group is derived from one end of each 'extender' acetate, and so the rationale for the term 'polyketide' will be apparent from Figure 7.1.

A number of polyketides are shown in Figure 7.2, together with the biosynthetic origins of their backbone structures. With the exception of OTC, which has the unusual three-carbon starter unit, the others are composed solely of acetate, i.e., acetate is the starter unit common to the other three. Each skeleton is derived from a different number of acetate extenders: for tetracenomycin, 9 acetates are added to the acetate starter making 10 in total; actinorhodin has 7 extenders (8 in total); and frenolicin has 8 extenders (9 in total). Thus, one source of variation in this family of polyketides is the numbers of extender units added.

From Figure 7.2, it will be apparent also that these different polyketides are folded and cyclized in different ways. OTC and tetracenomycin are folded to produce a 'tetracyclic' structure. The juxtapositions of the keto groups predicate how the cyclization takes place to form the four-ring structure. In the case of actinorhodin and frenolicin (Figure 7.2), when two keto groups find themselves close together (as in the right-hand rings) then one of the oxygen atoms actually gets incorporated into the ring (as an ether).

Each of these four antibiotics contains a six-carbon aromatic ring, and often they are given the generic name 'aromatic polyketides'. Figure 7.2 illustrates

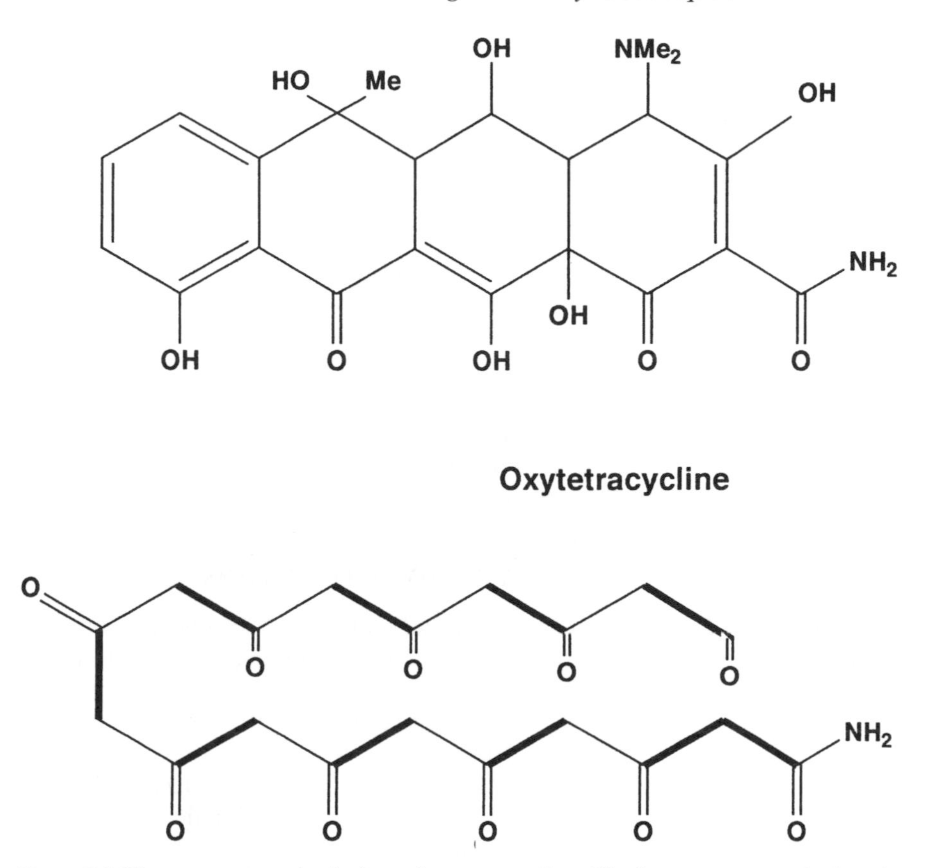

Figure 7.1 The upper structure is that of oxytetracycline. The lower structure depicts the biosynthetic origin of the backbone, from a three-carbon 'starter' unit (shown as an inverted bold V) and eight additions of a two-carbon 'extender' acetate unit (each shown as a bold line, with an imaginary carbon atom at each end).

that variation within the family of aromatic polyketides can arise through choice of starter unit, different numbers of acetate 'extender' units added, and/or different folding and cyclization patterns.

7.4 MOLECULAR GENETICS OF AROMATIC POLYKETIDE BIOSYNTHESIS

The analysis of the OTC biosynthetic genes began, as described above, with the cloning of a gene for resistance to OTC. In fact, two resistance genes were cloned. The first, *otrA* (for oxytetracycline resistance), renders the streptomycete ribosome resistant to the action of OTC (normally OTC interferes with the ribosome to arrest translation and kill the cell). The second, *otrB*, is responsible

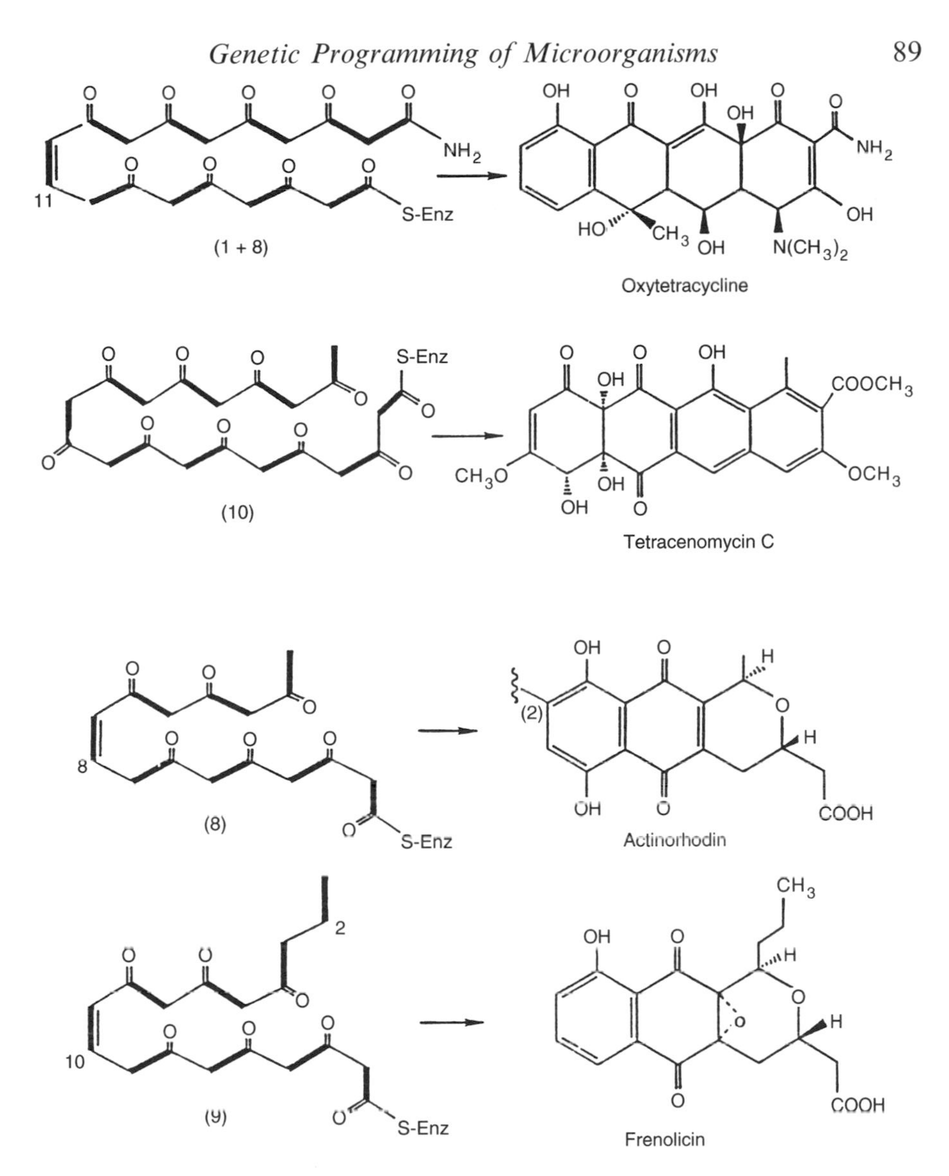

Figure 7.2 Biosynthetic origin of four polyketides, showing differences in starter unit (for OTC), number of extender units (although all extenders are acetate) and folding/cyclization patterns. This diagram has been modified from Khosla *et al.* (1993).

for active export of OTC from the cell (so that OTC never has an opportunity to bind to the intracellular ribosomes). The two resistance genes are located close to each other on the chromosome (Figure 7.3), and between them the genes for biosynthesis are located. There are around 25 genes responsible for the biosynthesis of OTC, but they can be grouped conveniently into those associated with the biosynthesis of the backbone skeleton and those which

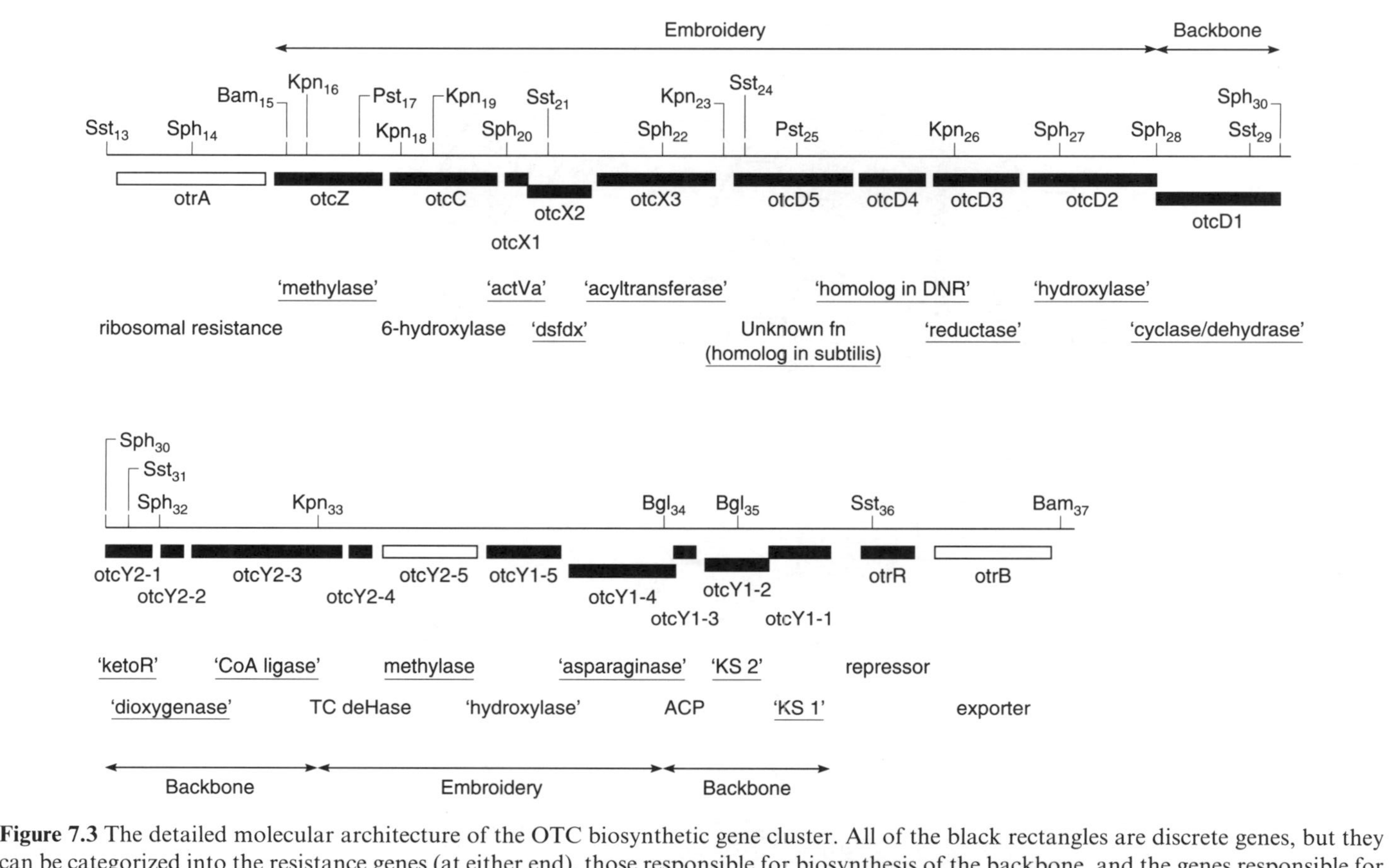

Figure 7.3 The detailed molecular architecture of the OTC biosynthetic gene cluster. All of the black rectangles are discrete genes, but they can be categorized into the resistance genes (at either end), those responsible for biosynthesis of the backbone, and the genes responsible for adding further chemical groups such as methyls and hydroxyls (collectively and colloquially termed 'embroidery' in this diagram).

are involved in subsequent addition of methyl and hydroxyl groups, which are tacked on as 'embroidery' after the backbone has been made (Figure 7.3).

The gene clusters for biosynthesis of tetracenomycin and actinorhodin have been cloned and analysed extensively by the laboratories of Professor R. Hutchinson (Madison, WI) and Professor D. Hopwood (Norwich, UK). The sub-cluster of genes which determine the backbone structure of frenolicin has also been analysed in Norwich. Taken together with the OTC cluster, all four gene clusters share similarity with each other in some regions (presumably because these gene products are undertaking the similar reactions of polyketide assembly) whereas each cluster also has some unique genes (which presumably impart unique features on each structure).

There are three particular genes in each cluster which are involved in the addition of extender units to the growing polyketide chain (Figure 7.4). They are found in the same order in each cluster. This high degree of conservation provided the ideal experimental 'test bed' for investigating the factors that determine the length of polyketide chain made.

The small acyl carrier protein (ACP, Figure 7.4) is the anchor to which the growing polyketide chain is attached. ACP proteins have a serine residue conserved at the active site which forms the chemical linkage between the ACP protein and the growing chain. It is thought that the ACP acts as a molecular pivot, rotating the growing polyketide chain into the active sites of the enzymes undertaking the biochemical reactions of polyketide assembly. It therefore seemed reasonable that the identity of the ACP might confer the specificity for length of chain. By carefully undertaking some molecular biological needlepoint, Chaitan Khosla (while working in the Norwich lab) began to exchange ACP genes between the different gene clusters (Khosla *et al.*, 1993). The rationale here was that, if the tetracenomycin ACP was plugged into the gap left after the actinorhodin ACP was cut out of that cluster AND

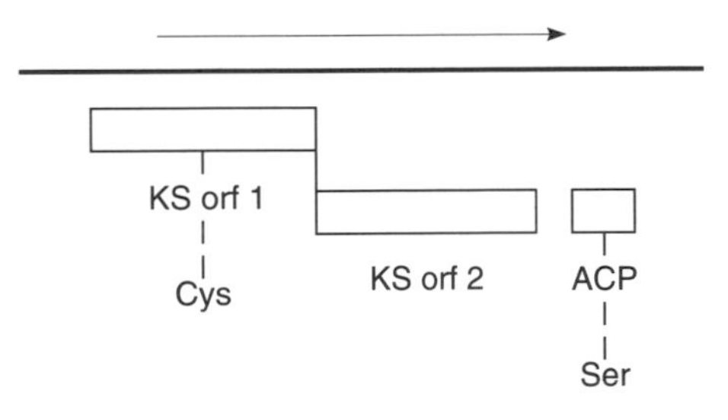

Figure 7.4 The molecular architecture of the three genes responsible for polyketide chain assembly: keto-synthase, (KS orf 1) responsible for catalysis through its active-site cysteine; KS orf 2, now known to specify the length of chain made; and the acyl carrier protein (ACP) to which the growing polyketide chain is attached via the active-site serine. The arrow shows the direction of transcription.

this hybrid cluster now made a twenty-carbon structure rather than the sixteen-carbon molecule made by the intact actinorhodin cluster, then this would be good evidence that the ACP specified chain length. It did not.

Attention was then focused on the two genes (orf 1 and orf 2) just upstream from the ACP (Figure 7.4). They had a fascinating architecture: orf 1 and orf 2 were nearly identical and linked together through an overlap in their genes, designed to ensure that equal amounts of each protein were made. The major difference between the two proteins was that orf 1 contained the catalytic cysteine required to undertake the biochemical reaction that adds an extender unit to a growing polyketide chain, whereas orf 2 did not. By exchanging genes between clusters, it was deduced that the orf 2 was the major determinant of chain length. It interacts with any orf 1 partner to instruct it on how many extender units have to be added. Orfs 1 and 2 have now been renamed. Orf 1 is now termed 'KS/AT' (keto-synthase, acyltransferase) to describe its enzymatic function in chain assembly, and orf 2 has been renamed 'CLF' (chain length formation determinant). The first part of this molecular detective story had been solved (for more details, see Hunter and Hill, 1997).

Among the four structures shown in Figure 7.2, OTC is unique in having the unusual three-carbon starter unit. Inspection of the OTC gene cluster (Figure 7.3) in relation to the other three (not shown) reveals that the OTC cluster has a unique 'acyltransferase' gene whereas the others have none. The anticancer drug daunorubicin also uses a three-carbon starter unit, and its gene cluster also contains an acyltransferase. Thus, there is a good correlation between the presence of this gene in any cluster and the structure of the polyketide chain made by that cluster containing an unusual starter unit. The acyltransferase probably loads the unusual starter unit on to the polyketide synthase. Then, in the absence of any specific acyltransferase, the default situation is that an acetate starter unit will be used. Therefore, at present we have a good indication of what determines the specificity of the starter unit, but definitive proof will require further experimentation.

The folding and cyclization of polyketide chains appears to be determined by enzymes collectively termed 'cyclases'. Within the OTC cluster, a cyclase gene can be identified clearly (Figure 7.3). Polyketide chains are highly reactive, due to their alternate keto groups. The natural tendency is for some keto groups to react (as shown in Figure 7.1 for OTC) to form carbon–carbon bonds which result in the formation of ring structures. It is believed that different cyclases contain 'pockets' into which growing polyketide chains can bind, and the way in which the chain is held (in 3-dimensional space) by a particular cyclase will govern which carbon–carbon bonds are made and hence the ring structure of the backbone. More work is needed to establish the ground rules for cyclization: this is not an easy topic to study as the chemistry is quite difficult.

A set of outline 'design rules' has been deduced to rationalize how the genes for biosynthesis of aromatic polyketide backbones determine the structure that is made (McDaniel *et al.*, 1995). Novel gene clusters have been created by bringing together component genes from a number of polyketide clusters, including the four mentioned above (Figure 7.2) (McDaniel *et al.*, 1994). These combinations of genes have never been seen before in nature and (hardly surprisingly), when they were introduced into a *Streptomyces* strain, novel metabolites not previously seen in nature were made (Figure 7.5). Some of these compounds have mild antibiotic activity (Fu and Khosla, 1995). Of course, these biological activities may be modified and enhanced by further chemical or biochemical modification of the basic structures. Such 'semi-synthetic' approaches have been a major strategy of the pharmaceutical industry over the years, and this new recombinant approach provides a series of new 'synthons' to plug into that well oiled machine. Also these molecules need to be

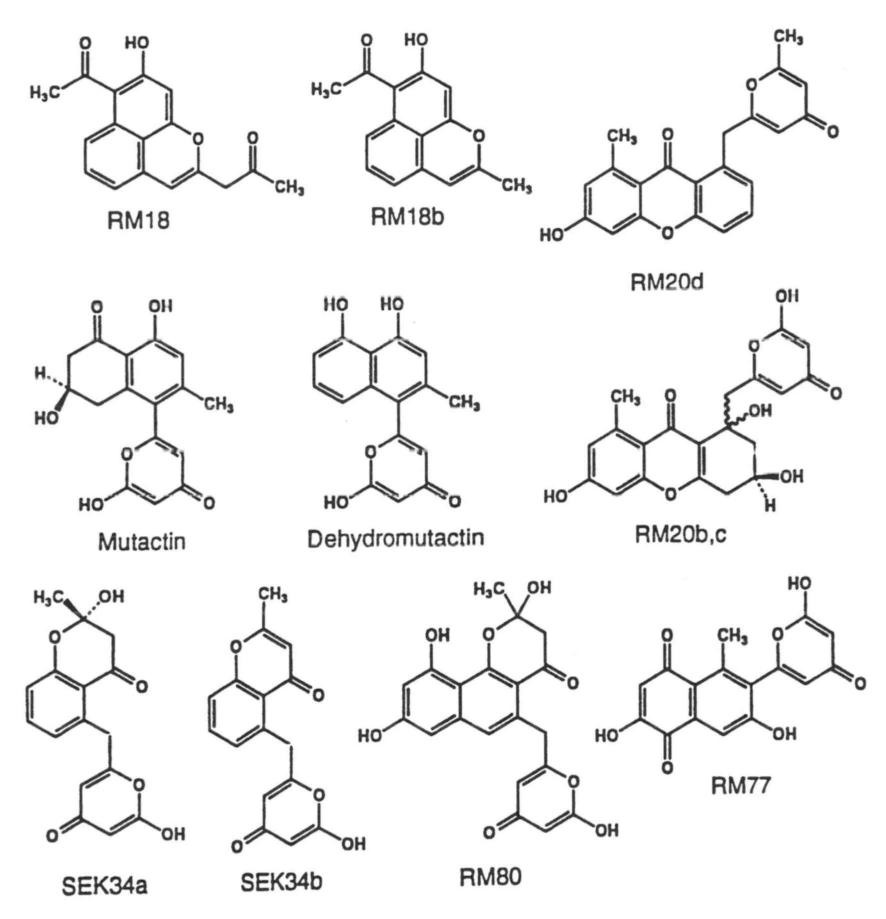

Figure 7.5 An assortment of novel polyketide structures made through creation of recombinant polyketide synthase gene clusters. The majority of these structures are described in McDaniel *et al.* (1994).

tested in high-throughput screens for potential utility as therapeutic agents, rather than antibiotics *per se*.

7.5 USING SINGLE GENES FROM ANTIBIOTIC CLUSTERS AS A MOLECULAR TOOLBOX

Although the emphasis of this chapter has been how the backbones of aromatic polyketides are derived, the gene clusters also contain a vast array of genes which add functionality to the backbone structures (the 'embroidery genes' of Figure 7.2). As well as the hydroxylation, amination, and methylation functions encoded in the OTC cluster, glycosylation also is a common feature used to embroider polyketide backbones. By mixing and matching single genes into novel clusters, a large degree of stuctural diversity can be generated in a stereospecific manner. A single hydroxylation gene was shown (Shen and Hutchinson, 1994) to be able to hydroxylate the polyketide backbone of tetracenomycin in three positions, demonstrating the potential of this approach.

7.6 FUTURE PERSPECTIVES FOR THE 'COMBI-GEN' APPROACH

Substantial efforts have been made by the pharmaceutical industry to capitalize on the recently developed combinatorial chemistry approach to deriving new chemical entities (see Chapter 9 by Shuttleworth), in which chemical building units are assembled on a chemical matrix to derive molecular diversity. The molecular biology approach described here presents an alternative (but not mutually exclusive) strategy. Combinatorial chemistry and 'combi-gen' should be undertaken in parallel to maximize the chance of discovering new drugs.

The rich experience of the antibiotics and fermentation industry teaches us that the small titres (less than 100 mg/l) of these novel chemical entities (such as shown in Figure 7.5) produced today by 'combi-gen' can be improved substantially in the future. The natural antibiotics such as tetracycline and penicillin were produced only at 10 mg levels when they were first discovered, but the fermentation industry now produces them at the level of many tens of grams per litre. Thus, the fermentation industry is ready and able to commercialize these new products as soon as proof of biological function is demonstrated.

Although this chapter has limited its scope to aromatic polyketides, the molecular biology of biosynthesis of a second (non-aromatic) class of polyketides, typified by the macrolide structure of erythromycin, is now well developed (e.g., Staunton *et al.*, 1996). This second class of polyketide is made by the same chemistry as described here, but the enzymes (and genes) are modular in structure, allowing opportunities to mix and match with even greater diversity than the aromatic polyketides.

Genome sequencing programs, including the Human Genome Sequencing Program, are now operating. A number of microbial genomes have been sequenced already and there are plans to sequence around fifty more. These new data are providing valuable insights into the roles of microbes as pathogens. However, an added spin-off is the recognition that many polyketide synthase genes are lurking within the genomes of these microbes. The mycobacteria are a particularly rich source of novel polyketide genes, immediately recognizable by searching the databases. We need not limit ourselves to the *Streptomyces* when isolating genes to construct novel combinations. However, it is likely that the *Streptomyces* will be used as the recipients for these novel gene combinations: their reputation as cell factories, able to produce large quantities of metabolites, is unsurpassed.

REFERENCES

Fu, H. and Khosla, C. (1995) Antibiotic activity of polyketide products derived from combinatorial biosynthesis: implications for directed evolution, *Molecular Diversity*, **1**, 121–4.

Hopwood, D.A., Bibb, M.J., Chater, K.F., Kieser, T., Bruton, C.J., Kieser, H.M., Lydiate, D.J., Smith, C.P., Ward, J.M. and Schrempf, H. (1985) *Genetic Manipulation of Streptomyces – a Laboratory Manual*, John Innes Foundation.

Hunter, I.S. and Hill, R.A. (1997) Tetracyclines: chemistry and molecular genetics of their formation, in *Biotechnology of Industrial Antibiotics* (ed. W.R. Strohl), Marcel Dekker, New York.

Khosla, C., McDaniel, R., Ebertkhosla, S., Torres, R., Sherman, D.H., Bibb, M.J. and Hopwood, D.A. (1993). Genetic construction and functional-analysis of hybrid polyketide synthases containing heterologous acyl carrier proteins, *Journal of Bacteriology*, **175**, 2197–204.

McDaniel, R., Ebert-Khosla, S., Hopwood, D.A. and Khosla, C. (1995) Rational design of aromatic polyketide natural products by recombinant assembly of enzymatic subunits, *Nature*, **375**, 548–54.

McDaniel, R., Ebert-Khosla, S., Fu, H., Hopwood, D.A. and Khosla, C. (1994) Engineered biosynthesis of novel polyketides: influence of downstream enzyme on the catalytic specificity of a minimal aromatic polyketide synthase, *Proceedings of the National Academy of Sciences of the USA*, **91**, 11542–6.

Shen, B. and Hutchinson, C.R. (1994) Triple hydroxylation of tetracenomycin-A2 to tetracenomycin-C in *Streptomyces glaucescens* – characterisation of the tcmG gene in *Streptomyces lividans* and characterisation of the tetracenomycin-A2 oxygenase, *Journal of Biological Chemistry*, **269**, 30726–33.

Staunton, J.S., Caffrey, P., Aparicio, G.F., Roberts, G.A. Bethell, S.S. and Leadlay, P.F. (1996) Evidence for a double-helical structure for modular polyketide synthases, *Nature Structural Biology*, **3**, 188–92.

8

Small Molecule Drug Screening Based on Surface Plasmon Resonance

Sanj Kumar and Kerstin Gunnarsson

8.1 INTRODUCTION

Understanding of a disease, identification of its therapeutic target and development of an approved pharmaceutical product is both a time-consuming and expensive process. Therefore, strong efforts are made to quicken the advance of drug candidates from hit to product. Biomolecular interaction analysis (BIA) is one of the emerging new technologies used in drug discovery, and has the potential to ensure that the evaluation of hits is not going to be the rate-limiting step in the drug discovery process. BIA measures interactions between two or more molecules by immobilizing one of the interactants (in BIA terminology this component is always called the 'ligand') on the surface of a sensor chip and then passing a solution containing the other interactant (the 'analyte') over this surface under controlled flow conditions. BIA is a general method with respect to interactant identity, and can be applied to all molecules expressing affinity for each other, such as biomolecules (proteins, nucleic acids, carbohydrates or lipids), low molecular weight compounds (signalling substances, pharmaceuticals, vitamins, pesticides, etc), or larger particles (vesicles, viruses, bacteria, cells). The specificity of the interaction analysis is determined by the identity and biological activity of the immobilized ligand. The interaction process is followed in real time without the need for any labels such as fluorescent tags or radioactivity. Data from an analysis provide information as to whether components bind to each other or not, and the kinetic constants and affinity of the interaction. The concentration of analyte in solution may also be calculated. Often, the analyte does not even need to be purified, but can be studied directly in crude extracts, coloured samples, fermentation broths, serum, etc. Multiple analyses can be performed on the same surface, which shortens

Advances in Drug Discovery Techniques. Edited by Alan L. Harvey
ISBN 0 471 97509 5

the total experimental procedure considerably and keeps the consumption of precious ligand compounds to a minimum.

8.2 PRINCIPLES OF BIA

BIA is a mass-sensitive technology based on surface plasmon resonance (SPR) detection (Jönsson and Malmqvist 1992), which monitors changes in the refractive index of the solution close to the surface of a sensor chip (Figure 8.1). The refractive index is related directly to the mass concentration in the surface layer, and increases when analyte binds to a ligand immobilized on a sensor chip. BIA experiments are performed under continuous flow conditions using a designed microfluidic system for the delivery of a sample to the sensor surface. The ligand is covalently bound or affinity captured to, e.g., a dextran matrix present on the sensor chip which preserves the conformation of the molecule.

When analyte is injected over the sensor chip, any change in mass concentration on the surface is detected as an SPR response expressed in resonance

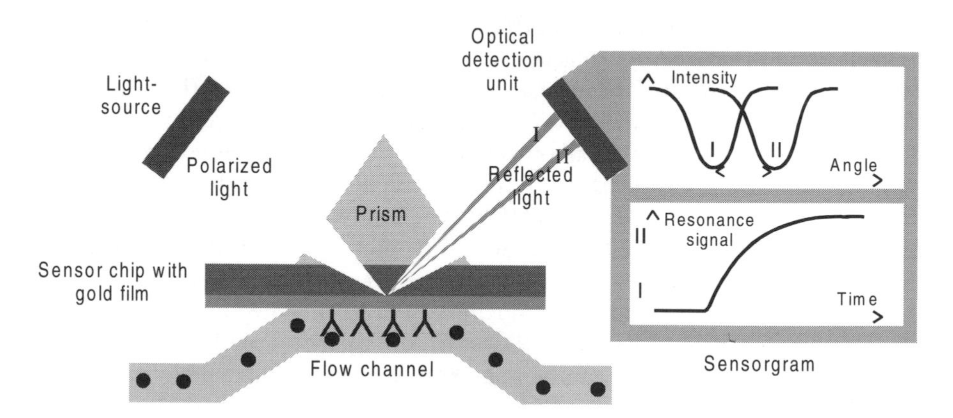

Figure 8.1 Biacore systems employ an optical phenomenon called surface plasmon resonance (SPR) as a detection system. SPR occurs when light illuminates the interface of two components each of a different refractive index and between which a metal film has been inserted. An evanescent wave created by total internal reflection interacts with free electron clouds in the metal causing a drop in the intensity of the reflected light. The angle (resonance angle) at which this reflectance minimum occurs is sensitive to the refractive index near the metal surface. The light source is a high efficiency LED with a wavelength near the infrared region. A wedge of light is directed towards the sensor surface and the reflected light is monitored with a two-dimensional diode array detector. The metal film properties (gold), the wavelength, and the refractive index of the denser medium (glass) are kept constant. Therefore the resonance angle will vary only with change in refractive index caused by changes in mass as molecules accumulate in the aqueous medium. When this change in resonance angle (reflectance minimum) is plotted against time(s) a sensorgram is created.

units. The continuous display of RU as a function of time is referred to as a sensorgram (Figure 8.2). The sensorgram provides a considerable amount of information. First, it shows that a binding interaction is occurring. In addition, it is possible to calculate the association and dissociation rate constants. The affinity of an interaction can be calculated from these constants or measured under steady state conditions.

When one interaction cycle is complete, the surface may be regenerated using conditions that remove bound analyte without affecting the activity of the immobilized ligand.

The preparation of a surface typically requires 100 μl of a 5–100 μg/ml ligand solution. For each analyte injection, typically 50–100 μl of solution is used (the required concentration is dependent on the affinity constant for the interaction investigated). As an example, Takemoto, Skehel and Wiley (1996) developed an SPR assay for the binding of monovalent influenza virus haemagglutinin to its sialic acid receptor. A complete inhibition curve could be obtained using less than 0.5 mg of inhibitor with an assumed K_D in the mM range.

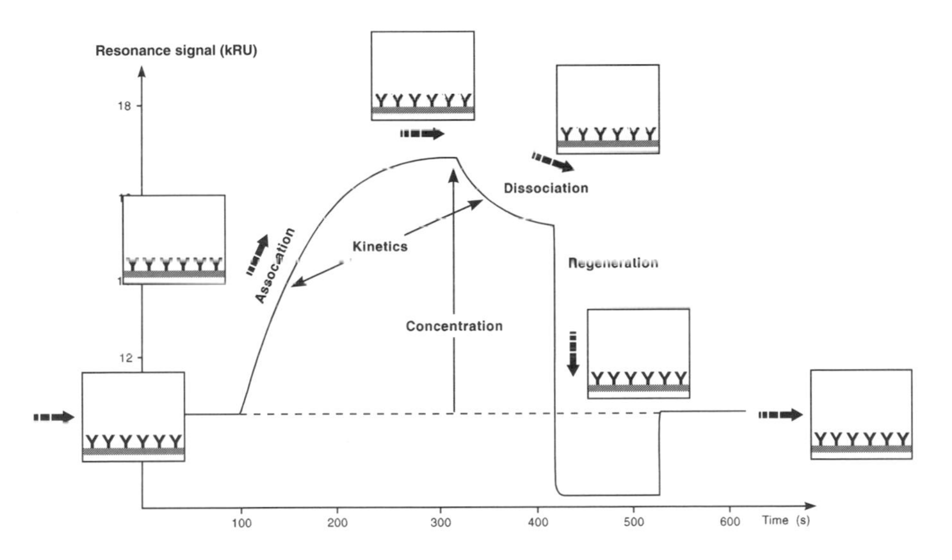

Figure 8.2 A typical sensorgram of an antibody–antibody interaction. The baseline (buffer flow) in this sensorgram is the covalently immobilized antibody. As the interacting antibody is injected the progress of the binding is monitored as a change in mass concentration on the surface. The sensorgram gives information about the association of the complex which can be followed. The equilibrium or steady state is displayed readily and when the analyte (antibody) is replaced by buffer flow the dissociation of the complex is monitored. The amount bound reflects the concentration of the sample. The introduction of a regenerated pulse (e.g., HCl) washes away all bound material and the surface is ready for a new injection.

Since the analyte does not need to be purified, BIA is a suitable method for screening purposes. BIA experiments are quick to design and carry out. A typical binding curve with association and dissociation is taken up in 5–10 min. Immobilization and optimization of the experiment normally is performed in less than one day.

8.2.1 Detection principle of surface plasmon resonance

Surface plasmon resonance is an optical phenomenon that occurs in thin metal films when light is reflected from the surface. When photons from a beam of polarized light are focused on a thin metal surface within a certain angle interval, a fraction of the light energy may be absorbed by the metal owing to interaction with the delocalized electrons (plasmons), causing the reflection to decrease or disappear (Figure 8.1). The specific angle of incidence is a function of the refractive index of the medium on and in the vicinity of the non-illuminated side of the metal film. For Biacore®, these are the sensor chip surface and the liquid sample cell. The refractive index is directly correlated to the mass concentration of solute in the surface layer. The detection is thus sensitive to mass concentration changes and may be used to measure the adsorption of molecules in a label-free manner. The resonance angle or signal is expressed in resonance units (RU). A response (i.e., a change in resonance signal) of 1000 RU corresponds to a change in surface concentration on the sensor chip of about 1 ng/mm^2 for proteins.

8.2.2 Sensor chip surfaces

The sensor chip (Figure 8.3) is a glass slide coated on one side with a thin gold film to which a surface coating is covalently attached (Löfås and Johnsson, 1990; Stenberg *et al.*, 1991). On most sensor chips the coating is a carboxylated dextran matrix which provides a hydrogel in which the ligand may be immobilized, giving a hydrophilic environment and preventing direct adsorption of proteins on to the gold surface. Furthermore, a matrix allows more molecules to interact per unit area as compared with a flat surface, leading to an increase in capacity. BIA differs from many surface interaction techniques in that the ligand/capturing molecule may be regenerated *in situ* and used repeatedly for up to 50–100 analyses or more (Blikstad *et al.*, 1996; MacKenzie *et al.*, 1997). Regeneration removes non-covalently bound molecules (analytes and/or affinity-captured ligands) from the surface. Suitable conditions depend on the nature of the ligand–analyte interaction and the ligand itself. The limiting factor when optimizing the regeneration procedure is the stability of the ligand, since the surfaces themselves are very stable to both high and low pH, detergents, denaturing reagents, and high salt concentrations.

The first step when preparing for an analysis is to immobilize one of the interacting species. To provide maximum flexibility in this step, a series of

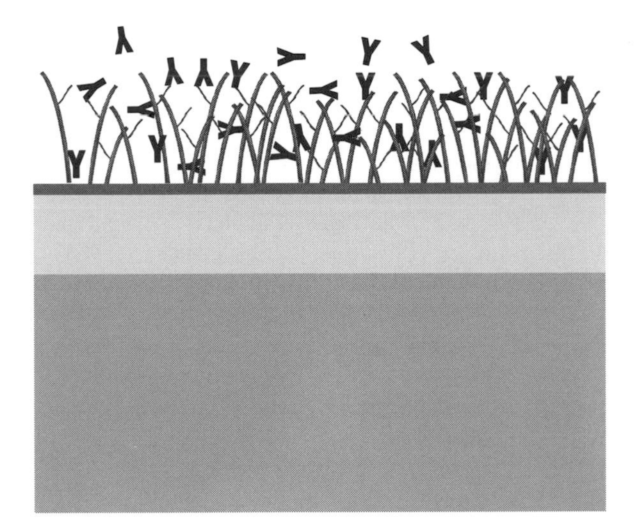

Figure 8.3 The sensor chip surface consists of a glass support on which a 50 nM thick gold film has been deposited. The gold layer has been covered with a non-crosslinked carboxylated dextran hydrogel coupled through an alkyl thiol linker layer. The dextran matrix (thickness 100 nM) increases surface capacity and discourages nonspecific binding. The carboxyl groups provide a chemical handle for covalent coupling of biomolecules.

surfaces and surface chemistries has been developed (see Table 8.1 and Figure 8.4) (Johnsson, Löfås and Lindquist, 1991; Löfås *et al.*, 1995). The most common method is amine coupling using carbodiimide chemistry to activate the carboxymethyl groups on the surface, which then react readily with uncharged primary amino groups on the ligand, resulting in a stable covalent attachment. Other chemistries for ligand immobilization include thiol–disulfide exchange and hydrazine-mediated coupling of aldehyde groups. Thiol exchange is especially attractive if the ligand has a single free cysteine, giving a well defined attachment site. The potential of this chemistry is enhanced by the possibility of including single cysteines in proteins by site-directed mutagenesis (Cunningham and Wells, 1993; Raghavan, Wang and Bjorkman, 1995).

An alternative to direct immobilization is ligand capturing, where an immobilized capturing molecule (e.g., streptavidin or a specific antibody) binds the ligand (e.g., biotinylated nucleic acid or a specific antigen). The affinity of the capturing interaction must be high to minimize dissociation of ligand during the analysis procedure. With capturing techniques, also the ligands may be adsorbed from crude mixtures. For example, immobilized RAMFc may be used to capture monoclonal antibodies from unfractionated hybridoma supernatants. Usually, the capturing molecule approach results in a well defined orientation of the ligand. Regeneration is typically performed back to the

Table 8.1 Sensor chips

Sensor chip	*Immobilization strategy*	*Functionality*
Sensor chip CM5	Covalent coupling of ligands via their amino, thiol, aldehyde or carboxyl groups	COOH
Sensor chip SA	Capture of biotinylated ligands	Streptavidin
Sensor chip NTA	Capture of histidine-tagged ligands	NTA metal chelate
Sensor chip HPA	Liposome-mediated hydrophobic adsorption of lipid layer or coating	Flat hydrophobic surface

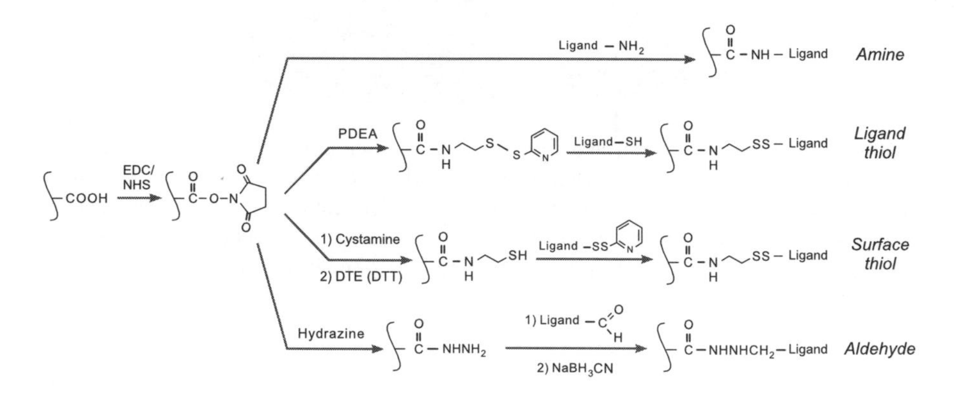

Figure 8.4 The most common coupling procedure is amine coupling, which couples molecules covalently through their primary amines. In cases where coupling through primary amines is unsuitable other alternatives are available, such as two approaches for thiol disulfide exchange. In addition, biotinylated reagents (e.g., peptides, DNA) may be immobilized with high affinity through amine coupled avidin surfaces. Molecules oxidized to aldehydes can couple to surfaces activated by hydrazine or carbohydrazide. The method is suitable for coupling of glycoconjugates and polysaccharides. Not shown are the many varied protocols where the first of the binding partners is captured to the chip surface by a covalently bound molecule, e.g., a high affinity antibody. Once the first molecule is captured, the interacting partner may then be injected and the binding profile studied.

capturing molecule. Since new ligand is captured for each analysis cycle, capturing techniques are advantageous in cases where the ligand would not tolerate the conditions required to dissociate the analyte.

Another type of surface coating is used on sensor chip HPA, which is covered with a hydrophobic layer allowing adsorption of lipid monolayers with the hydrophilic heads facing the solution. Ligands may be incorporated into or adsorbed onto the lipid monolayer, allowing studies of interactions in a membrane-like environment. Ligands with hydrophobic parts also may be adsorbed directly onto the HPA surface.

8.2.3 Microfluidic system

For interaction measurements, a flow system is superior to cuvette-based or macrofluidic systems for distribution of the analyte to the sensor surface and detector since fresh solution is continuously passing over the ligand surface. In Biacore® instruments, the integrated microfluidic cartridge (IFC) contains flow channels, sample loops and pneumatic valves, and controls delivery of sample and eluent to the sensor chip surface (Sjölander and Urbaniczky, 1991). A significant advantage with microfluidics is the rapid exchange of sample and buffer at the sensor surface, and the ability to control the contact time. All of these features are essential for high accuracy, precision and reproducibility. Flow cells for detection are formed by pressing the IFC against the sensor chip, keeping the dead volume between the sample loops and sensor chip surface to a minimum. The flow cells are designed for minimal sample dispersion and efficient mass transport of biomolecules to the sensor chip surface. No washing steps are needed to replace sample with buffer, which is crucial for the study of fast reaction kinetics.

To allow also low concentration samples to be handled without significant adsorption, the material in the flow system has a high chemical resistance and is protein compatible. Analyte concentrations down to 1 pM may be used.

With the multispot measurement facility available in the Biacore®2000 (Karlsson and Ståhlberg, 1995), it is possible to direct solution to different flow cells, allowing for a comparison with a 'blank' flow cell. Particularly for low response levels, it is important to have the possibility to correct for bulk refractive index effects with an in-line reference. Sensorgram detection may be performed in all four flow cells simultaneously, making it possible to improve kinetic and concentration measurements by covering a range of surface ligand concentrations and contact times in one analysis. By immobilizing different ligands in each flow cell multiple targets may be analysed, saving time when large numbers of samples are to be screened.

8.3 WHAT CAN BIA MEASURE?

A sensorgram will provide immediate information about whether two components interact with each other or not. This information may be used to screen and rank a series of analytes with respect to yes/no binding. Further analysis will also provide the interactants' affinity and kinetic constants. Usually, only minor changes in experimental design are required for determining these parameters with high accuracy.

Since the introduction on the market in 1990, BIA has been used extensively in basic research, leading to about 700 peer reviewed scientific publications by mid-1997. Many areas in contemporary medical and life science research are covered, such as: signal transduction including receptor–ligand interactions; immune regulation; molecular biology (nucleic acid–protein or nucleic acid–nucleic acid interactions); ligand 'fishing' (identify binders to a receptor); molecular recognition, e.g., cell adhesion, infectious mechanisms; enzyme reactions; molecular engineering; and target definition.

8.3.1 Interaction kinetics and affinity

The sensorgrams contain detailed kinetic information, since the monitoring of an interaction procedure is continuous and in real time. While kinetic analysis earlier was restricted to systems with favourable spectral properties, BIA technology has made studies on kinetic properties of most macromolecular interactions routinely accessible. The elucidation of kinetic parameters and comparison with other methods have been covered extensively in the literature (Myszka, 1997). The real-time data can be fitted to theoretical models allowing rate constants to be calculated. Association rate constants in the range 10^3–$10^7/M^{-1}\,s^{-1}$ (Shen, Hage and Sebald, 1996) and dissociation rate constants in the range 10^{-1}–$10^{-5}/s^{-1}$ (van der Merwe *et al.*, 1993; Khilko *et al.*, 1995) can be determined and affinities in the 10^{-4}–10^{-12} M range (van der Merwe *et al.*, 1993; Schier *et al.*, 1996).

For screening purposes the qualitative information obtained from an overlay plot often is sufficient to rank candidates, and there may be no need to quantify data in terms of rate and affinity constants at this stage of the investigation (Schier *et al.*, 1996).

An example of qualitative ranking is given in Figure 8.5. Three different monoclonal antibodies directed towards the HIV protein p24 were immobilized and the same concentration of analyte, p24, was injected over the different surfaces. When p24 bound to immobilized MAb 18 or 28, steady state binding was obtained during the time of the experiment. Since MAb 18 gave a higher steady state response, it could be concluded immediately that MAb 18 has higher affinity for p24 than has MAb 28. With MAb 1 immobilized, the interaction time was too short for a steady state to be reached, and a direct affinity comparison with this MAb was not possible. However, it is evident that MAb 1 interacted more slowly with p24 than did MAb 18 and 28. Data from the dissociation phase showed that p24 dissociated fairly rapidly from MAb 28 and comparatively slowly from MAbs 1 and 18.

8.3.2 Binding specificity

The most intuitive application of real-time BIA is the qualitative investigation of binding specificity: do two components interact with each other or not? This is used for several types of investigation briefly discussed now.

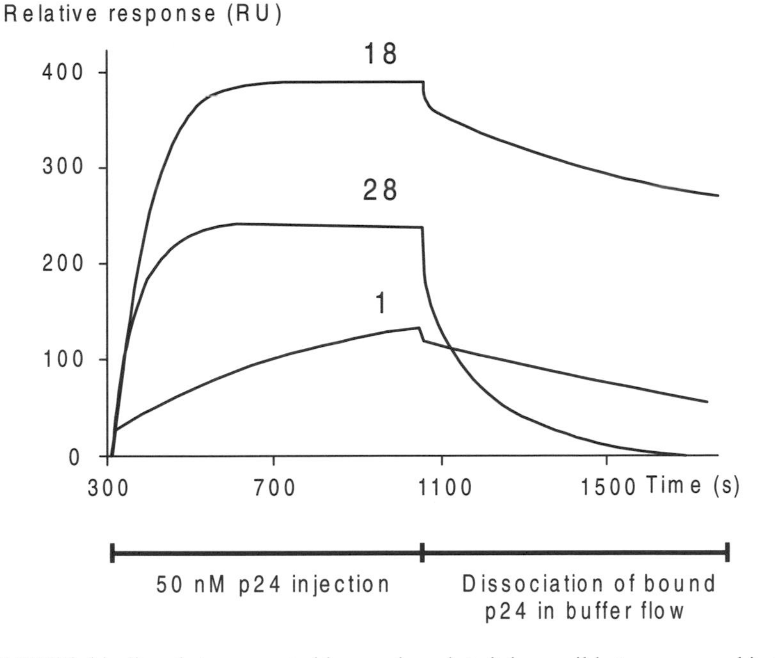

Figure 8.5 With binding data presented in overlay plots it is possible to compare binding levels, association rates and dissociation rates for different analytes reacting with the same binding partner. If the same concentration of analyte is used it is possible also to compare binding curves for one analyte interacting with different immobilized partners. This figure compares the binding of p24 antigen to three different MAbs. When p24 binds to immobilized MAb 18 or 28, steady state binding is obtained. Since MAb 18 gives a higher steady state response we can immediately conclude that MAb 18 has a higher affinity for p24 than MAb 28. With MAb 1 immobilized the interaction time is too short for steady state to be reached and a direct affinity comparison with this MAb is not possible. What we can say is that MAb 1 interacts more slowly with p24 than do MAb 18 and 28. Looking at the dissociation phase we find that p24 dissociates fairly rapidly from MAb 28 and more slowly from MAbs1 and 18. For screening purposes often the qualitative information obtained from an overlay plot is sufficient, and there may be no need to quantitate data in terms of rate and affinity constants.

(a) Screening for binding partners

An important application area for BIA is screening for binding partners to target molecules, e.g., hybridoma cultures for monoclonal antibody production, natural product extracts and combinatorial libraries for potential new therapeutics. Molecules of interest can be traced rapidly in crude preparations. During the search for better binders (which is the case in the search for both agonists and antagonists), ranking with respect to kinetic constants will help to identify the most interesting leads for further investigation. Usually, the dissociation rate is expected to decrease for a better binder, since the association rate is largely diffusion-limited (Cunningham and Wells, 1993; Schier *et al.*, 1996).

BIA has been used to screen cell culture supernatants for ligands which bind to the extracellular domain of the 'orphan' receptor ECK protein-tyrosine kinase (Bartley *et al.*, 1994). The extracellular domain of ECK was immobilized on the sensor chip surface, and binding response in serum-depleted cell culture supernatants was measured after 25s injections of supernatants. A binding activity was found in several distinct eukaryotic cell lines and not in others, indicating that within the protein mixtures that comprise these supernatants a specific ECK receptor binding molecule is present. After purification protocols and further studies, the ligand to ECK receptor was determined to be B61, a molecule already known to be involved in cellular defence response pathways.

(b) Binding site mapping

By using a set of antibodies, the relative positions of binding sites can be determined in a pairwise approach by the fact that two antibodies either interfere or are independent in their binding. This binding site mapping gives useful information for, e.g., the construction of sandwich assays, where it is necessary that the first and second antibodies bind to different epitopes (Johne, Gadnell and Hansen, 1993).

In binding site mapping, every binding step in the procedure is monitored, giving control information about the amounts of first antibody and antigen bound. These data are not obtained from label-dependent techniques, which usually allow only the last step to be monitored. With BIA it is possible also to test the pair of antibodies in reversed order. No washing steps which could remove interactants with a high dissociation rate constant are needed since the association phase is observed directly. This ensures that low-affinity interactions can be detected also.

(c) Multi-component binding studies

Since every stage in a binding sequence can be monitored directly and in real time with BIA it is possible to follow the formation of multimolecular complexes. Schuster *et al.* (1993), for example, have used BIA to study the formation of a quaternary signalling complex from the chemotactic system of *E. coli*. They found that a stable quaternary complex could be formed on the sensor chip surface only when the components were injected in a particular order, and other combinations gave no complex. This experiment highlights the advantage of being able to see the effect of each individual injection, as opposed to using an end-point measurement.

8.3.3 Concentration assay

Reliable concentration measurements can be made of complex mixtures such as serum, fermentation broth, crude cell extracts and cell suspensions. By the very nature of this technology, concentration determinations with BIA often are less susceptible to interference by other components in the sample than are conventional techniques such as ELISA. With a suitable ligand, the functionally active concentration is determined, as distinct from the total concentration.

By controlling the amount of immobilized ligand, it is possible to measure concentrations over a broad range, typically 10^{-10}–10^{-3} M (Strandh *et al.*, 1996). The working range of a particular assay is determined by the affinity of the interactant for the analyte. Concentration assays can be constructed either to measure concentrations directly or by an inhibition assay (Biacore Application Note AN 304).

The amount of bound analyte is measured after a fixed injection time. By injection of a secondary interactant in a sandwich assay design, the sensitivity can be enhanced further. Accurate measurements can be made also in systems where binding does not approach equilibrium, and in this case, a longer sample injection will increase the sensitivity of the assay.

8.3.4 Low molecular weight analytes

The sensitivity of the instrumentation in combination with the high binding capacity of polymer matrix surfaces makes the binding of a low molecular weight (200–1000 Da) analyte to immobilized ligands directly detectable (Hendrix *et al.*, 1997), thus allowing screening of synthetic compounds or natural extract libraries. The system is stable in the presence of up to 8% DMSO. By using in-line references to compensate for refractive index bulk effects, reliable measurements may be carried out in such mixed solvents (Karlsson and Fält, 1997).

8.3.5 Interfacing with other techniques

As a step in the further identification and quantification of material that adsorbs onto or subsequently is eluted from the BIA sensor chip, mass spectrometry (MS) can be integrated readily. In the case of adsorbed material, the sensor chip itself can be prepared directly in the solid phase sample matrix for MALDI MS. Eluted material can be analysed both with MALDI and electrospray ionization (ESI) (Sönksen *et al.*, 1997).

8.4 A TOOL IN DRUG DISCOVERY

BIA can provide the user with a great amount of relevant information to be used throughout the whole drug discovery process (see Figure 8.6), facilitating

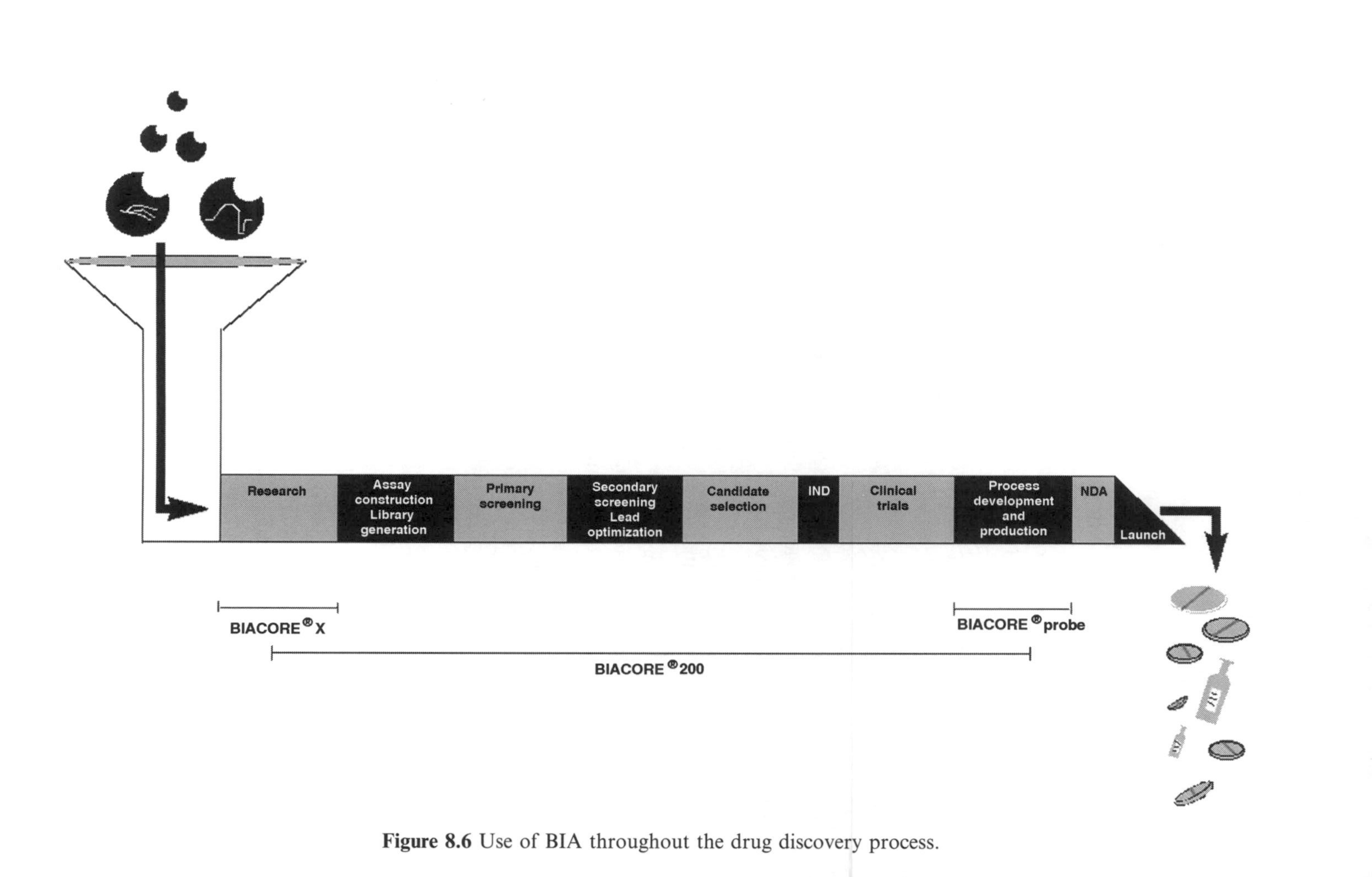

Figure 8.6 Use of BIA throughout the drug discovery process.

transitions between the different phases and saving efforts in method development, since the same assay can be used during many phases of the process.

Drug discovery begins with the elucidation of a particular biochemical pathway or receptor and the identification of a compound that will modulate this course of events. Basic research aiming at an understanding of a disease and the mechanism behind it involves a variety of scientific disciplines and is fundamental to the development of new drugs. BIA is used in programs for many diseases, as outlined in Table 8.2.

In the validation phase, BIA can facilitate the design and set-up of different control experiments. Very small changes in binding affinity can be studied, which is a prerequisite for analysing the functional effect and thermodynamic implications of minor structural changes in interacting molecules. BIA has been used to study the kinetic effects of alanine substitution scanning through 30 residues in the structural binding domain of hGH (Cunningham and Wells, 1993). It was shown that the side-chains of roughly 25% of these residues account for the major part of the binding energy. Thus the functional domain is significantly smaller than the structural domain, implying that it might be possible to design smaller hormone mimics for therapeutic purposes.

BIA has proved useful in the development of antibodies for different purposes, for instance as applied in immunoassay development. Antibodies with optimal epitope recognition and binding properties can be selected rapidly, with substantial time savings compared to other methods (Johne, Gadnell and Hansen, 1993).

When developing radio-labelled antibodies intended for cancer treatment it is important to reach a high distribution ratio between the target and the surroundings quickly in order to minimize total body radiation. One approach to a rapid clearance of non-targeted circulating antibodies (primary antibody, PA) is to administer a secondary antibody (SA). By using BIA, the kinetic

Table 8.2 Use of BIA in research programs for various diseases

Disease	*Reference*
Inflammation	Arend *et al.* (1994)
Cancer	Schier *et al.* (1996)
Autoimmune diseases	Moosmayer *et al.* (1996)
Diabetes	Kennedy, Skillen and Self (1994)
Heart diseases	Roos *et al.* (1995)
Allergy	Meng, Singh and Wong (1996)
AIDS	VanCott *et al.* (1995)
Alzheimer's disease	Zambrano *et al.* (1997)
Antibiotic resistance	Demeule *et al.* (1995)

constants between antigen and PA can be determined together with that between PA and SA. The PA/SA ratio administered can be adjusted to such kinetic information (Ullén *et al.*, 1996).

8.4.1 From hits. . .

BIA can determine quickly how well a drug molecule will bind to a particular biological target or, conversely, if molecules will not bind together. Ranking normally does not require any sophisticated experimental set-up, and thus the information obtained with BIA gives a broad foundation for making decisions about which leads should be pursued further.

Another benefit is that the interaction process is continuously monitored, as opposed to the end-point result obtained with ELISA assays. With end-point analyses it is difficult to tell if a negative is true or false, but with real-time analysis, it is immediately apparent if one of the components of the assay is defective.

BIA can be used in a multi-channel mode to construct control surfaces where different receptors are immobilized in each flow cell. Compounds that exhibit activity exclusively against the primary target probably possess the necessary selectivity and are less likely to be generally toxic (Broach and Thorner, 1996).

The power of BIA in primary screening is demonstrated in the work of Shane Taremi and coworkers at the Schering-Plough Research Institute, NJ (Taremi *et al.*, 1996). The team has developed a high-throughput assay for low molecular weight cytokine antagonists acting at IL-4 receptor using a Biacore®2000. Their protocol involves the IL-4 receptor being immobilized to the chip surface, then a mixture of cytokine and test compound is injected over the surface to detect any binding. The screening was made in a mixed solvent (2% DMSO) and 2000 samples were tested in less than 24 hours to provide the team with valuable information on the mechanisms of action of various cytokine inhibitors. Appropriate controls, such as control IL-4, IL-4 with DMSO or IL-4 plus the free receptors, were included. In this example, four positives were found, i.e., components which prevented IL-4 from binding to the receptor. The assay was consistent from one end of a microtitre plate to the other.

The design of a BIA screening assay allows many variations. First, there is a choice of which component should be immobilized. Second, the analyte determination could be made either by a direct binding assay and/or by a competitive assay. Furthermore, the desired level of information from the screening assay, ranging from binding/no binding (yes/no) to affinity ranking, or true kinetic/equilibrium data, determines the degree of optimization necessary.

If HTS screening exceeding the capacity of the equipment is required, BIA can support the development of other assays, e.g., ELISA, and with BIA the combinations of monoclonal antibodies suitable for an assay will be apparent immediately (Bennett *et al.*, 1995).

BIA also has supported other screening protocols successfully, such as the novel TNF-α dual lanthanide label assay (DELFIA), developed at Xenova Discovery, Slough, UK (patented) (see Chapter 3 by Hill). In this case DELFIA uses time resolved fluorescence (TRF) to detect potent specific microbial inhibitors of p55 receptor/ligand interactions via screening of their NatChem™ libraries. Potential inhibitors identified from this assay have been evaluated further using BIA to determine the effect of primary hits on TNF-α receptor binding (Sohal *et al.*, 1997).

The strategy of using the dual label TRF screening assay to reduce false positives and the rapid secondary analysis data generation afforded by BIA has reduced timelines and costs drastically in the identification of novel compounds of interest.

8.4.2 . . . to a developed compound

The optimization of a drug lead that emerges from primary screening into a useful drug is an iterative process, and involves both pharmacokinetic and biophysical characterization. During this phase, BIA is excellent for more thorough kinetic studies and for specificity studies where the optimized leads can be screened for binding activity against different receptors. Such studies provide a more critical selection of leads that will be candidates for resource-demanding animal testing, and ultimately lead to a developed compound.

Since a serum can be studied without prior purification, BIA is suitable also for clinical assays. BIA has been used, for example, to test the response to HIV vaccines based upon a subunit of a recombinant HIV envelope protein. The humoral immune response in both HIV-infected and uninfected volunteers who had been immunized was examined (VanCott *et al.*, 1995).

8.5 CONCLUSIONS

The fact that drug development times have become longer and increasingly expensive is due in part to a shift of focus from drugs used to treat acute infectious diseases towards a spectrum of drugs used to treat relatively complex chronic degenerative diseases. At the same time, the pharmaceutical industry is developing new insights into the molecular mechanisms of disease that will lead to a deeper understanding of the relationship between molecular cause and clinical effect. As the roles of multi-gene disorders are better understood, it is possible that diseases that are seemingly unrelated on a clinical level may well be closely related on the molecular level. In this light it is clear that new ways to strengthen data obtained in the drug discovery process will be of crucial importance in the efforts to reduce the time lag from the original idea to a final approval of a drug. In order to rationalize pharmaceutical development, new approaches to drug discovery systematically using knowledge of molecular function must be explored and implemented. Since innovative core technologies that address these needs both qualitatively and quantitatively will be critical for success, we can expect BIA to play a central role in this context.

REFERENCES

Arend, W.P., Malyak, M., Smith, M.F., Whisenand, T.D., Slack, J.L., Sims, J.E., Giri, J.G. and Dower, S.K. (1994) Binding of IL-1 alpha, IL-1 beta, and IL-1 receptor antagonist by soluble IL-1 receptors and levels of soluble IL-1 receptors in synovial fluids, *The Journal of Immunology*, **153**, 4766–74.

Bartley, T.D., Hunt, R.W., Welcher, A.A., Boyle, W.J., Parker, V.P., Lindberg, R.A., Hsieng, S.L., Colombero, A.M., Elliott, R.L., Guthrie, B.A., Hoist, P.L., Skrine, J.D., Toso, R.J., Zhang, M., Fernandez, E., Trail, G., Varnum, B., Yarden, Y., Hunter, T. and Fox, G.M. (1994) B61 is a ligand for the ECK receptor protein-tyrosine kinase, *Nature*, **368**, 558–60.

Bennett, D., Morton, T., Breen, A., Hertzberg, R., Cusimano, D., Appelbaum, E., McDonnell, P., Young, P., Matico, R. and Chaiken, I. (1995) Kinetic characterization of the interaction of biotinylated human interleukin 5 with an Fc chimera of its receptor alpha subunit and development of an ELISA screening assay using real-time interaction biosensor analysis, *Journal of Molecular Recognition*, **8**, 52–8.

Blikstad, I., Fägerstam, L.G., Bhikhabhai, R. and Lindblom, H. (1996) Detection and characterization of oligosaccharides in column effluents using surface plasmon resonance, *Analytical Biochemistry*, **233**, 42–9.

Broach, J.R. and Thorner, J. (1996) High-throughput screening for drug discovery, *Nature*, **384**, 14–6.

Cunningham, B.C. and Wells, J.A. (1993) Comparison of a structural and functional epitope, *Journal of Molecular Biology*, **234**, 554–63.

Demeule, M., Vachon, V., Delisle, M.-C., Beaulieu, E., Avril-Bates, D., Murphy, G.F. and Béliveau, R. (1995) Molecular study of P-glycoprotein in multidrug resistance using surface plasmon resonance, *Analytical Biochemistry*, **230**, 239–47.

Hendrix, M., Priestly, E.S., Joyce, G.F. and Wong, C.-H. (1997) Direct observation of aminoglycoside-RNA interactions by surface plasmon resonance, *Journal of the American Chemical Society*, **119**, 3641–8.

Johne, B., Gadnell, M. and Hansen, K. (1993) Epitope mapping and binding kinetics of monoclonal antibodies studied by real time biospecific interaction analysis using surface plasmon resonance, *Journal of Immunological Methods*, **183**, 191–8.

Johnsson, B., Löfås, S., and Lindquist, G. (1991) Immobilization of proteins to a carboxymethyldextran-modified gold surface for biospecific interaction analysis in surface plasmon resonance sensors, *Analytical Biochemistry*, **198**, 268–77.

Jönsson, U. and Malmqvist, M. (1992) Real time biospecific interaction analysis. The integration of surface plasmon resonance detection, general biospecific interface chemistry and microfluidics into one analytical system, in *Advances in Biosensors*, Vol. 2, JAI Press, London, pp. 291–336.

Karlsson, R. and Fält, A. (1997) Experimental design for kinetic analysis of protein–protein interactions with surface plasmon resonance biosensors, *Journal of Immunological Methods*, **200**, 121–33.

Karlsson, R. and Ståhlberg, R. (1995) Surface plasmon resonance detection and multi-spot sensing for direct monitoring of interactions involving low-molecular-weight analytes and for determination of low affinities, *Analytical Biochemistry*, **228**, 274–80.

Kennedy, D.M., Skillen, A.W. and Self, C.H. (1994) Glycation of monoclonal antibodies impairs their ability to bind antigen, *Clinical and Experimental Immunology*, **98**, 245–51.

Khilko, S.N., Jelonek, M.T., Corr, M., Boyd, L.F., Bothwell, A.L.M. and Margulies, D.H. (1995) Measuring interactions of MHC class I molecules using surface plasmon resonance, *Journal of Immunological Methods*, **183**, 77–94.

Krone, J.R., Nelson, R.W., Dogruel, D., Williams, P. and Granzow, R. (1997) BIA/MS: interfacing biomolecular interaction analysis with mass spectroscopy, *Analytical Biochemistry*, **244**, 124–132.

Löfås, S. and Johnsson, B. (1990) A novel hydrogel matrix on gold surfaces in surface plasmon resonance sensors for fast and efficient covalent immobilization of ligands, *Journal of the Chemical Society, Chemical Communications*, 1526–8.

Löfås, S., Johnsson, B., Edström, Å., Hansson, A., Lindquist, G., Müller Hillgren, R.-M. and Stigh, L. (1995) Methods for site controlled coupling to carboxymethyldextran surfaces in surface plasmon resonance sensors, *Biosensors & Bioelectronics*, **10**, 813–22.

MacKenzie, C.R., Hirama, T., Lee, K.K., Altman, E. and Young, N.M. (1997) Quantitative analysis of bacterial toxin affinity and specificity for glycolipid receptors by surface plasmon resonance, *Journal of Biological Chemistry*, **272**, 5533–8.

Meng, Y.G., Singh, N. and Wong, W.L. (1996) Binding of cynomolgus monkey IgE to a humanized anti-human IgE antibody and human high affinity IgE receptor, *Molecular Immunology*, **33**, 635–42.

Moosmayer, D., Wajant, H., Gerlach, E., Schmidt, M., Brocks, B. and Pfizenmaier, K. (1996) *Journal of Interferon and Cytokine Research*, **16**, 471–7.

Myszka, D.G. (1997) Kinetic analysis of macromolecular interactions using surface plasmon resonance, *Current Opinions in Biotechnology*, **8**, 50–7.

Raghavan, M., Wang, Y. and Bjorkman, P.J. (1995) Effects of receptor dimerization on the interaction between the class I major histocompatibility complex-related Fc receptor and IgG, *Proceedings of the National Academy of Sciences of the USA*, **92**, 11200–4.

Roos, W., Eymann, E., Symannek, M., Duppenthaler, J., Wodzig, K.W.H., Pelsers, M. and Glatz, J.F.C. (1995) Monoclonal antibodies to human heart fatty acid-binding protein, *Journal of Immunological Methods*, **183**, 149–53.

Schier, R., McCall, A., Adams, G.P., Marshall, K.W., Merritt, H., Yim, M., Crawford, R.S., Weiner, L.M., Marks, C. and Marks, J.D. (1996) Isolation of picomolar affinity anti-c-erbB-2 single-chain Fv by molecular evolution of the complementarity determining regions in the center of the antibody binding site, *Journal of Molecular Biology*, **263**, 551–67.

Schuster, S.C., Swanson, R.V., Alex, L.A., Bourret, R.B. and Simon, M.I. (1993) Assembly and function of a quaternary signal transduction complex monitored by surface plasmon resonance, *Nature*, **365**, 343–6.

Shen, B.-j., Hage, T. and Sebald, W. (1996) Global and local determinants for the kinetics of interleukin-4/interleukin 4 receptor α chain interaction, *European Journal of Biochemistry*, **240**, 252–61.

Sjölander, S. and Urbaniczky, C. (1991) Integrated fluid handling system for biomolecular interaction analysis, *Analytical Chemistry*, **263**, 2338–45.

Sönksen, C.P., Jansson, Ö., Malmqvist, M. and Roepstorff, P. (1997) An improved method for combining BIACORE® instruments with MALDI mass spectrometry, Poster presented at the 12th International Symposium on Affinity Interactions, Kalmar, 15–19 June, 1997.

Sohal, J., Elcock, C., MacAllan, D., Abery, J. and Hill, D. (1997) Poster presented at the TNF-Antagonists Meeting, organized by Cambridge Symposia.

Stenberg, E., Persson, B., Roos, H. and Urbaniczky, S. (1991) Quantitative determination of surface concentrations of protein with surface plasmon resonance using radiolabelled proteins, *Journal of Colloid and Interface Science*, **143**, 513–26.

Strandh, M., Persson, B., Stenhag, K., Nilshans, H. and Ohlson, S. (1996) The use of biomolecular interaction analysis for studies of weak affinity antibody–antigen recognition, in Report from the Sixth European BIAsymposium, Montpellier 21–23 October 1996, p. 160.

Takemoto, D.K., Skehel, J.J. and Wiley, D.C. (1996) A surface plasmon resonance assay for the binding of influenza virus hemagglutinin to its sialic acid receptor, *Virology*, **217**, 452–8.

Taremi, S.S., Prosise, W., Rajan, N., O'Donnell, R.A. and Le, H.V. (1996) Human interleukin-4 receptor complex: neutralization effect of two monoclonal antibodies, *Biochemistry*, **35**, 2322–2331.

Ullén, A., Riklund Åhlström, K., Hietala, S.-O., Nilsson, B., Ärlestig, L. and Stigbrand, T. (1996) Secondary antibodies as tools to improve tumor to non-tumor ratio at radioimmunolocalisation and radioimmunotherapy, *Acta Oncologica*, **35**, 281–5.

VanCott, T.C., Bethke, F.R., Burke, D.S., Redfield, R.R. and Birx, D.L. (1995) Lack of induction of antibodies specific for conserved, discontinuous epitopes of HIV-1 envelope glycoprotein by candidate AIDS vaccines, *Journal of Immunology*, **155**, 4100–10.

van der Merwe, P.A., Brown, M.H., Davis, S.J. and Barclay, A.N. (1993) Affinity and kinetic analysis of the interaction of the cell adhesion molecules rat CD2 and CD48, *The EMBO Journal*, **12**, 4945–54.

Zambrano, N., Buxbaum, J.D., Minopoli, G., Fiore, F., De Cadia, P., De Renzis, S., Faraonio, R., Sabo, S., Cheetham, J., Sudol, M. and Russo, T. (1997) Interaction of the phosphotyrosine interaction/phosphotyrosine binding-related domains of Fe65 with wild-type and mutant Alzheimer's β-amyloid precursor proteins, *Journal of Biological Chemistry*, **272**, 6399–6405.

9

An Overview of Combinatorial Chemistry and Its Applications to the Identification of Matrix Metalloproteinase Inhibitors (MMPIs)

Stephen J. Shuttleworth

9.1 COMBINATORIAL CHEMISTRY

9.1.1 Background

With the significant growth in molecular biology and advanced biochemistry, the number of lead compounds available for biological testing no longer meets the increasing number of available biological screens as illustrated in Figure 9.1. This has led to a 'bottleneck' within the area of lead discovery.

To a large extent the problem has been addressed by combinatorial chemistry, a technique which has been the subject of several extensive review articles in the chemical literature (Pavia, Sawyer and Moos, 1993; Gallop *et al.*, 1994; Gordon *et al.*, 1994; Nielsen, 1994; Doyle, 1995; Eichler and Houghten, 1995; Terrett *et al.*, 1995; Wilson and Czarnik, 1998). Combinatorial chemistry, in essence, involves the preparation of large numbers of compounds in a time- and resource-effective manner. The technique makes use of synthetic strategies which are designed to facilitate the preparation of large numbers of analogues, commonly referred to as libraries, under similar reaction conditions, either as mixtures or as individual compounds (Figure 9.2). For example, substrate A may be treated with reagants a, b, c, etc., leading to a large number of target compounds (Aa, Ab, Ac etc.) involving relatively few manipulations.

Combinatorial chemistry was pioneered in the mid-1980s by Geysen (Geysen, Meloen and Barteling, 1984; Geysen, Rodda and Mason, 1986;

Advances in Drud Discovery Techniques. Edited by Alan L. Harvey
ISBN 0 471 97509 5

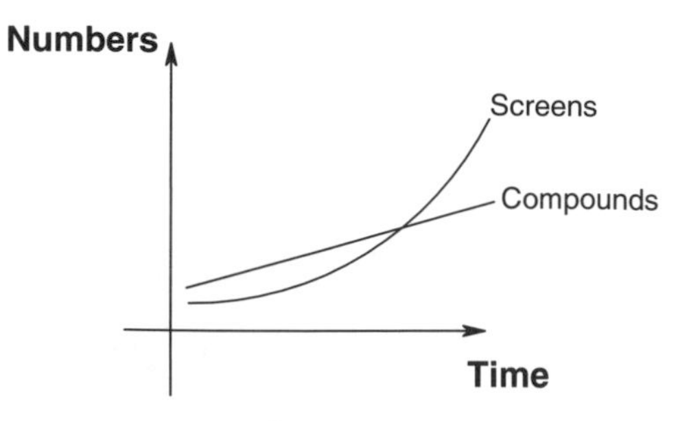

Figure 9.1 The 'new lead bottleneck': biological screens now outnumber available compounds.

Substrates		Reagents		Products
A		a		Aa Ab Ac Ad Ae etc ...
B		b		Ba Bb Bc Bd Be etc ...
C	+	c	⇨	Ca Cb Cc Cd Ce etc ...
D		d		Da Db Dc Dd De etc ...
E		e		Ea Eb Ec Ed Ee etc ...
etc ...		etc ...		

Figure 9.2 Simplistic representation of the concept of combinatorial synthesis.

Geysen *et al.*, 1987), and was subsequently developed further by Houghten (1985) and Furka *et al.* (1991). Initial studies by all three researchers focused on the synthesis of peptide libraries. However, the true potential of combinatorial chemistry, and its use in the identification of small-molecule, drug-like compounds has been fully realized only since the early part of this decade; this is perhaps typified by Bunin and Ellman (1992) whose report discusses the synthesis of a benzodiazepine analogue library. Since that time, combinatorial chemistry has become firmly established within the pharmaceutical industry as a means of expeditiously producing large numbers of compounds for drug discovery programmes; the breadth of chemistry employed for library synthesis has widened, and, accordingly, the technique has grown considerably (De Witt *et al.*, 1993; Chen *et al.*, 1994; Balkenhohl *et al.*, 1996; Thompson and Ellman, 1996).

The use of combinatorial synthesis in medicinal chemistry research has enabled the pharmaceutical industry to overcome the lead bottleneck problem, and several recent cases have been reported where hit compounds identified from within libraries have been selected as candidates for clinical trials (Borman, 1997). It should be noted that the combinatorial synthesis has been employed successfully also in the agrochemical industry (Parlow and Normansell, 1995).

9.1.2 Mixtures or single compounds?

As discussed above, combinatorial libraries may be prepared as mixtures of compounds or as single compounds in parallel (Lin and Shapiro, 1996; Lam, 1997). Until recently the former approach was favoured. However, owing to problems encountered with chemical analysis and deconvolution (determining the active component) of compound mixtures, coupled with the identification of 'false positives' (active mixtures whose theoretical components are less active or inactive upon discrete resythesis), the synthesis of large mixtures has become less desirable (Konings *et al.*, 1996; Wilson-Lingardo *et al.*, 1996; Pirrung, 1997). Consequently, many research groups now prepare compound libraries either as mixtures composed of small numbers of compounds, or as individual entities.

There remains substantial debate over which approach is the more acceptable. A compromise may be in compound 'pooling', where the targets are synthesized discretely and then judiciously combined to form decipherable mixtures. Compound pooling addresses the problems of analysis and deconvolution mentioned above, though not necessarily of false positives.

9.1.3 Techniques for combinatorial chemistry

Combinatorial chemistry may be performed using polymeric supports (resins), a method commonly referred to as solid phase synthesis (Merrifield, 1986; Maud, 1992), or in solution (Figure 9.3). More recently, the use of 'soluble' polymers (such as derivatized poly(ethylene glycol)) in the synthesis of combinatorial libraries also has been reported (Han and Janda, 1996; Wentworth, Vandersteen and Janda, 1997). This 'liquid phase' technique holds several well documented advantages over solid and solution phase chemistry, although it has not been investigated to nearly the same extent for combinatorial synthesis purposes.

To date, the most widely used technique has been solid phase chemistry. The theory behind solid phase chemistry is simple: an insoluble, polymeric support is anchored to a substrate via a linker group, and the construction of the desired compound may be effected using chemistries which do not compromise the stability of either the linker or the polymer. Following construction of the desired compound, the product is cleaved from the resin (Figure 9.4). The main advantage of this approach over solution chemistry is that reactions may be driven to completion by using vast excesses of reagents. Since the product is immobilized, these excess reagents may simply be washed away, allowing for products of high chemical purity to be released from the polymer. Solid phase synthesis is also easy to automate, and is firmly established for important areas in drug discovery such as peptide, (For review see Davies, 1992; for multiple peptide synthesis methods see *Angewandte Chemie International Edition in English),* **31**, 367–83; for semiautomated T-bag peptide synthesis

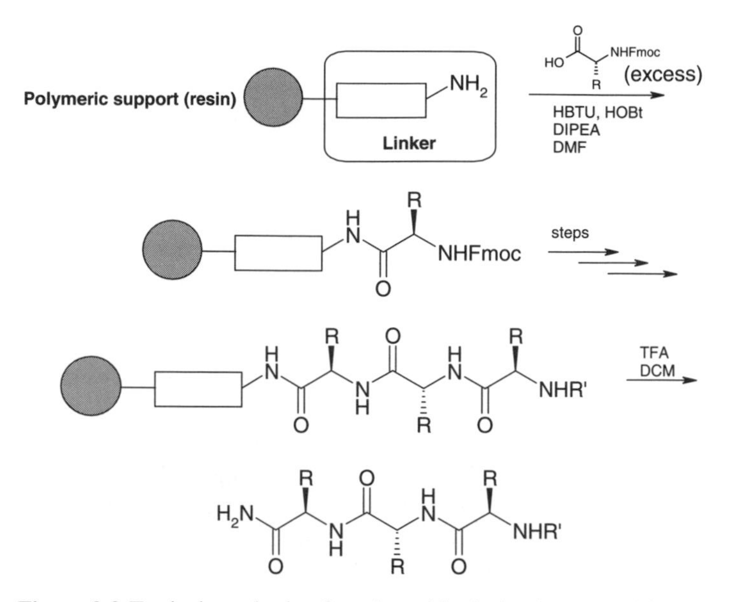

Figure 9.3 Typical synthesis of a tripeptide derivative on solid phase.

Solid Phase	Solution Phase
Range of chemistry growing	Wide range of chemistry
Analysis sophisticated	Analysis trivial
High product purity	Purity can be low
Easy to automate	Easy to automate
Amenable to many steps	Limited to two or three steps
Easy purification	Purification difficult
Limited quantities of product	Can prepare large quantities of material
Resin technology and expertise required	No new technology or expertise required
Resins and automation expensive	Costs not as high

Figure 9.4 Comparison of solution and solid phase chemistries.

using 9-fluoromethoxycarbonyl strategy see *Peptide Research*, **4**, 88–95; for recent studies see Tam and Lu, 1995; Emery *et al.*, 1996; O'Donnell, Zhou and Scott, 1996; Valero, Giralt and Andreu, 1996), oligonucleotide (For review see White, 1992; for recent studies see Wright *et al.*, 1993; Basu and Wickstrom, 1995; Le Bec and Wickstrom, 1996; McGuigan *et al.*, 1996) and carbohydrate synthesis. (For review see Sofia, 1996; for recent studies see Yan *et al.*, 1994; Adinolfi *et al.*, 1996; Boons, Heskamp and Hout 1996; Hunt and Roush, 1996, Shimizu *et al.*, 1996; Nicolaou *et al.*, 1997). Additionally the range of synthetic chemistry applicable to solid phase synthesis has increased markedly over recent years (Früchtel and Jung, 1996; Hermkens, Ottenheijm and Rees,

1996, 1997). The use of enzymes (Schulster *et al.*, 1994; Turner, 1996) chiral auxiliarics, and metal-promoted reactions (Deshpande, 1994; Hiroshige, Hauske and Zhou, 1995; Guiles, Johnson and Murray, 1996; Zhang and Maryanoff, 1997), all important in the synthesis of small–molecule targets, (Moon, Schore and Kurth, 1992, 1994; Allin and Shuttleworth, 1996; Burgess and Lim, 1997), is now commonplace in the field of resin chemistry. Solid phase heterocyclic chemistry has blossomed over the past year (Nefzi, Ostresh and Houghten, 1997), while the construction of polyamines (Nash, Bycroft and Chan, 1996) and biotinylated probes (Kumar *et al.*, 1994; Neuner, 1996) on resins has been reported. Solid phase synthesis of benzodiazepines (Bunin, Plunkett and Ellman, 1994; Boojamra, Burow and Ellman, 1995; Goff and Zuckerman, 1995; Plunkett and Ellman, 1995) and peptidomimetics (Virgilio and Ellman, 1994; Hoekstra *et al.*, 1996; Rotella, 1996) has received widespread coverage in the chemical literature. Diverse, small-molecule analogues have been assembled on solid phase using multi-component condensation reactions (Hamper, Dukesherer and South, 1996; Koh and Ellman, 1996; Mjalli, Sarshar and Baiga, 1996; Sarshar, Siev and Mjalli, 1996; Strocker *et al.*, 1996).

However, despite the merits of solid phase synthesis, there are some inherent drawbacks: resins are expensive, and reaction analysis is not trivial. The most commonly used analytical methods for monitoring the progress of solid phase reactions include mass spectrometry (Egner, Langley and Bradley 1995; Haskins *et al.*, 1995) infrared spectroscopy (Yan, Fell and Kumaravel, 1996; Chan *et al.*, 1997), and solid state (magic-angle spinning and gel-phase) NMR (Fitch, Detre and Holmes, 1994; Look *et al.*, 1994; Anderson *et al.*, 1996), all of which use sophisticated and expensive equipment. Further, being a niche technique, solid phase synthesis requires expertise in resin technology and an understanding of polymer science in order to be exploited successfully.

Solution phase combinatorial synthesis has been performed to great effect by several research groups (Smith *et al.*, 1994; Carell *et al.*, 1995; Holmes *et al.*, 1995; Merritt, 1995; Pirrung and Chen, 1995; Bailey, 1996; Boger *et al.*, 1996, 1997; Cheng *et al.*, 1996; Peterson, 1996; Shuttleworth, 1996b; Storer, 1996). The benefits of solution phase synthesis to combinatorial chemistry include the unlimited range of synthetic transformations which may be performed, its familiarity to the chemist, its low cost of automation, and its ease of analysis. The main disadvantage of this approach is that, invariably, low yields, and hence low purities, of products are obtained, making solution chemistry less useful for multi-step reactions for which, by contrast, solid phase synthesis is very well suited. To help address this problem, the use of polymer-supported scavengers (Kaldor *et al.*, 1996) and ion-exchange slurrying (Parlow, 1996; Gayo and Suto, 1997b; Siegel *et al.*, 1997) for purification of solution phase combinatorial synthesis has been developed. Indeed, the use of polymer-supported reagents in combinatorial solution synthesis is also under increasing investigation (Shuttleworth *et al.*, 1997).

In order to continue to develop as a viable technique for medicinal chemistry, combinatorial synthesis has relied upon the input of cutting edge technology from a wide range of sources including automation (Yan *et al.*, 1994), resin linker technology (Han, Walker and Young, 1996; Brown *et al.*, 1997; Gayo and Suto, 1997a; Uozumi, Danjo and Hayashi, 1997), encoding (for library deconvolution) (Nestler, Bartlett and Still, 1994; Baldwin *et al.*, 1995), and even micro-electronics (Nicolaou *et al.*, 1995). Combinatorial chemistry remains very much a developing area, and its future success is reliant upon a broad platform of scientific expertise.

9.1.4 Impact on drug discovery

Combinatorial chemistry impacts at two areas in the drug discovery process, namely lead generation and lead optimization (Figure 9.5). For lead generation, a large number of suitably diverse compounds are synthesized and then tested against the biological targets of choice. The chemical strategy employed for lead generation is required to be flexible and should employ a broad range of building blocks. Achieving suitable diversity of compounds may be accomplished by drawing upon computational chemistry expertise (Cohen *et al.*, 1990), and many software packages are available to assist the combinatorial chemist for this purpose; indeed computational chemistry has become vitally important in the assessment of diversity in both monomers and library products (Cramer *et al.*, 1996; Ferguson *et al.*, 1996; Patterson *et al.*, 1996). Once several potent compounds have been identified from primary screens,

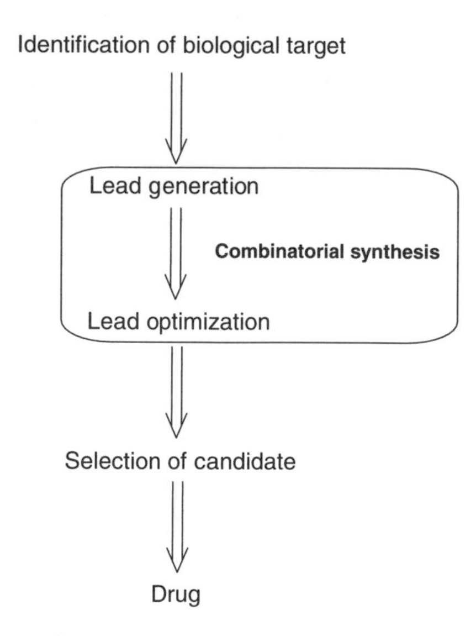

Figure 9.5 Impact of combinatorial chemistry in the drug discovery process.

these 'hits' must be developed further, and their physico-chemical properties and overall drug-like nature enhanced. Combinatorial chemistry plays a significant role at this lead optimization stage. More focused libraries are synthesized in this case, and specific templates, identified from primary screens, may be used to build streamlined sets of compounds. Owing to the nature of lead optimization, the combinatorial chemistry employed tends to be more focused than that used to generate lead compounds; specific building blocks are used, and single compounds often are prepared.

9.2 MATRIX METALLOPROTEINASES (MMPs)

9.2.1 MMPs

MMPs are a family of zinc-containing mammalian proteinases which have been the focus of considerable research over recent years (Beckett *et al.*, 1996; Davidson *et al.*, 1997). MMPs depend on zinc for catalysis, and on calcium for activity and stability. They are subclassed into four groups (Figure 9.6). The collagenases cleave triple-helical interstitial collagen, the stromelysins cleave proteoglycans, and the gelatinases cleave denatured collagen, elastin and types IV and V collagen. It is the ability of MMPs to cleave and degrade collagen and proteoglycan, two major components of connective tissue, which allows them to play an important role in the remodelling and degradation of the extracellular matrix of both basement membranes and connective tissue.

Collagenases

Interstitial collagenase (MMP–1);
Polymorphonuclear neutrophil (PMN)
Collagenase (MMP–8);
Collagenase-3 (MMP–13).

Stromelysins

Stromelysin-1, transin, proteoglycanase (MMP–3);
Stromelysin-2 (MMP–10); stromelysin 3 (MMP–11).

Gelatinases

Gelatinase A, 72 kDa-gelatinase, type IV Collagenase (MMP–2);
Gelatinase B, 92 kDa gelatinase (MMP–9).

Membrane–type matrix metalloproteinases

MT-1 (MMP–14)
MT-2 (MMP–15)
MT-3 (MMP–16)
MT-4 (MMP–17)

Figure 9.6 The matrix metalloproteinase family of enzymes.

9.2.2 Inhibition of MMPs

MMPs are secreted as inactive proenzymes. Upon their activation, MMPs are modulated by tissue inhibitors of metalloproteinases (TIMPs) (see Beckett *et al.*, 1996; references 1 and 2) and α_2-macroglobulin. Although a certain amount of activated MMPs may be necessary for normal matrix remodelling, excessive release of MMPs leads to connective tissue degradation and destruction.

Inhibition of MMPs has become an area of intense investigation for several years (Beckett *et al.*, 1996; Davidson *et al.*, 1997). Potent and selective inhibitors of MMPs have become desirable as potential therapeutic agents for controlling pathological processes in a range of diseases involving connective tissue damage. It is believed that inhibition of MMPs could be useful in the treatment of a wide range of inflammatory conditions (Han, Walker and Young, 1996; Brown *et al.*, 1997; Gayo and Suto, 1997a; Uozumi, Danjo and Hayashi, 1997) including rheumatoid arthritis, osteoarthritis, psoriasis, dermatitis, Alzheimer's disease and multiple sclerosis (Hughes *et al.*, 1995). Furthermore, inhibition of MMPs is believed to be effective in the treatment of other illnesses, such as cardiovascular disease and cancer (Hughes *et al.*, 1995).

Considerable effort has focused on MMP inhibitors (MMPIs) which are designed around the cleavage site of the enzyme and which contain a strong zinc-binding group. To date the majority of MMPI research has centred on the construction of substrate-based peptidic and peptidomimetic inhibitors possessing a hydroxamic acid unit as the zinc binding group. In such cases, enzyme recognition is provided by P_1', P_2', and in some cases P_3' interactions. The initial guide for substrate-based design was the glycine–isoleucine and glycine–leucine cleavage sites in the collagen molecules which are hydrolysed by collagenase. Until recently, compounds of this class have displayed poor oral activity in animal models, low bioavailability, and poor physico–chemical properties. Inhibitors possessing other zinc binding groups such as phosphinates, aminocarboxylates, and thiol-based ligands have, in general demonstrated reduced potency agains the MMPs.

Significant advances in the identification of MMP inhibitors have been made by researchers at several biotechnology and major pharmaceutical companies such as Celltech (Morphy *et al.*, 1994; Porter *et al.*, 1994), Sterling Winthrop (Gowravaram *et al.*, 1995; Singh *et al.*, 1995; Tomczuk *et al.*, 1995; Wahl *et al.*, 1995), Pfizer (Robinson *et al.*, 1996), Glaxo-Wellcome (Brown *et al.*, 1994), Parke-Davis (Ye *et al.*, 1994), Syntex (Castelhano *et al.*, 1995), Chiroscience (Baxter *et al.*, 1997), Merck and SmithKline Beecham (Bird *et al.*, 1995). Researchers at British Biotechnology (Dickens, Crimmin and Beckett 1994; Miller *et al.*, 1997), Roche (Broadhurst, 1993), and GlycoMed (Levy *et al.*, 1998), have made major discoveries in this area, with compounds **1** and **2** (Dickens, Crimmin and Beckett, 1994; Miller *et al.*, 1997), **3** (Broadhurst, 1993), and **4** (Levy *et al.*, 1998) (Figure 9.7).

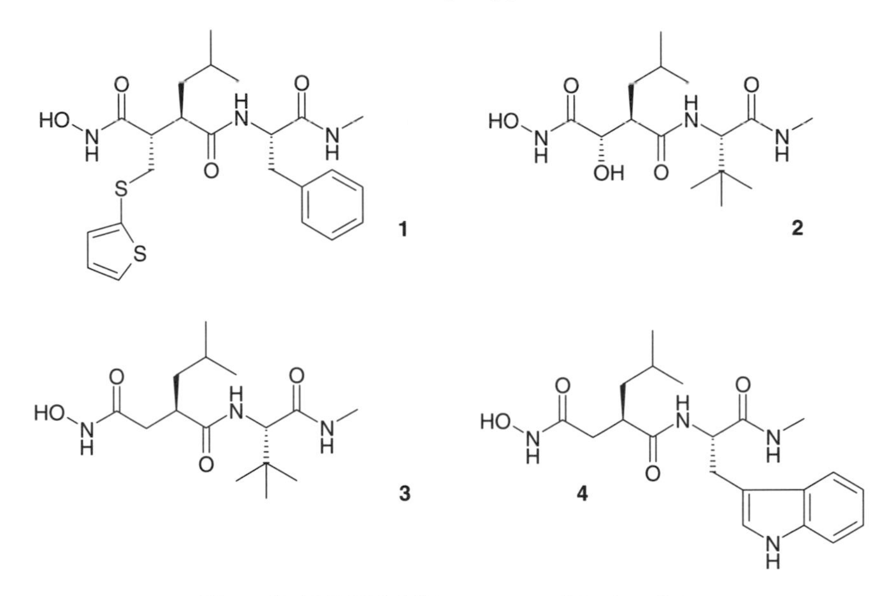

Figure 9.7 MMP inhibitors assessed in the clinic.

Other MMPIs recently identified are listed in Figure 9.8. These include the collagenase–selective, indolactam–based inhibitor **5** (Nuigel *et al.*, 1995), a stromelysin-active piperazic acid derivative **6** (Hughes *et al.*, 1995), and the broad-spectrum inhibitor **7** derived from Futoenone (Yeh *et al.*, 1995). Ciba-Geigy (**8**), Parke-Davis (**9**) and Bayer (**10**) have recently disclosed the structures of some of their non-substrate based MMPIs, which exhibit IC_{50}s of 17 nM (stromelysin), 150 nM (MMP–2), and 38 nM (MMP–2), respectively.

9.3 COMBINATORIAL SYNTHESIS OF MMPIs

As discussed above, considerable research has focused on inhibitors of MMPs since the early 1980s. However, *combinatorial* synthesis of MMPIs remains very much an emerging topic within medicinal chemistry research, and has not been the subject of extensive literature coverage. In this section, examples are discussed where combinatorial chemistry has been applied to the preparation of MMPI libraries. It is to be assumed that many research groups investigating MMPI science have employed combinatorial chemistry for their programmes; however, for the purpose of this section, only published material will be discussed.

At Chiroscience, potent MMP inhibitors have been identified (Scheme 9.1) from libraries of tripeptides prepared in solution (Shuttleworth, 1996a). Lead optimization of the core template **II** (Figure 9.9), identified in the Chiroscience laboratories (Baxter *et al.*, 1997), was studied using solution phase parallel

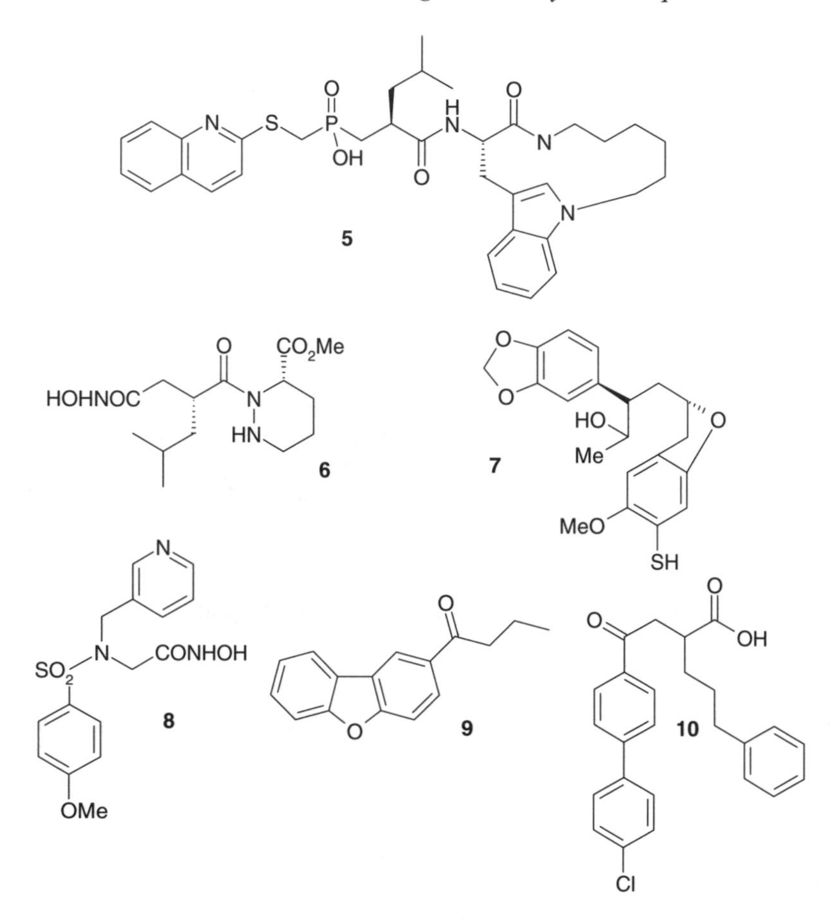

Figure 9.8 Examples of recently discovered MMPIs.

synthesis, with manipulation of the P_1', P_2', and P_3' positions of **11** being investigated using a range of amino acid derivatives. Compounds displaying IC_{50}s of 1 nM against gelatinase B were identified.

Researchers at Glaxo-Wellcome (Foley *et al.*, 1996) have investigated the solid phase synthesis of modified C and N terminal groups of the cysteine-derived lead compound **12** (Figure 9.10). The synthesis of two libraries composed of several hundred compounds of the type **12** was accomplished. The chemistry involved as shown in Scheme 9.2. In the first library, the phenylalanine residue was replaced with 50 amino acids. Heteroaromatic and aliphatic side chains of L-amino acids incorporated here led to compounds exhibiting collagenase (MMP–1) selectivity (Figure 9.11, **13**, **14**, and **15**). The second library (Figure 9.12) contained derivatives with the phenylalanine residue kept constant and the N terminus of the dipeptide capped with 150 different

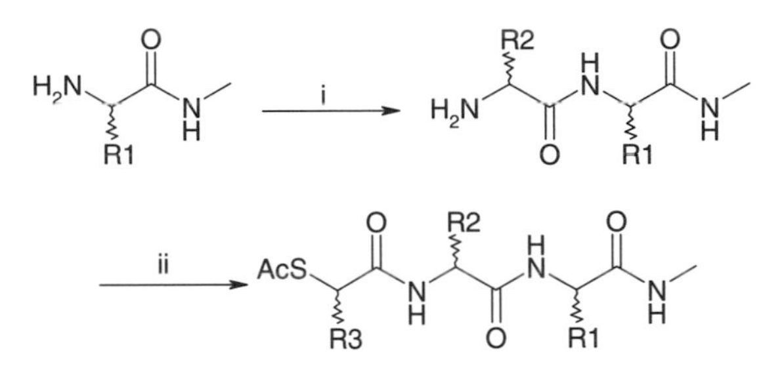

Scheme 9.1 Preparation of MMPIs described by Chiroscience using solution phase chemistry.

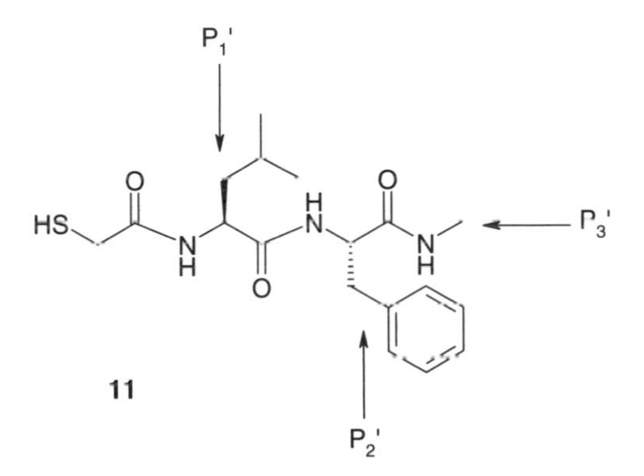

Figure 9.9 Thioamide derived MMP inhibitor discovered by Chiroscience.

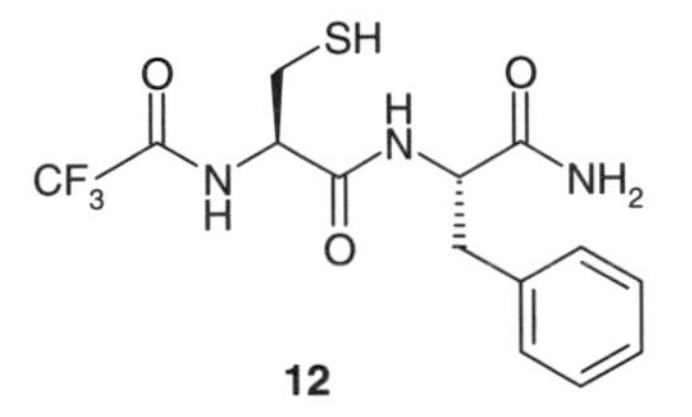

Figure 9.10 Collagenase-selective MMPI lead discovered by Foley *et al.* (1996). IC_{50}s: Gelatinase = >1000 nM; collagenase = 40 nM.

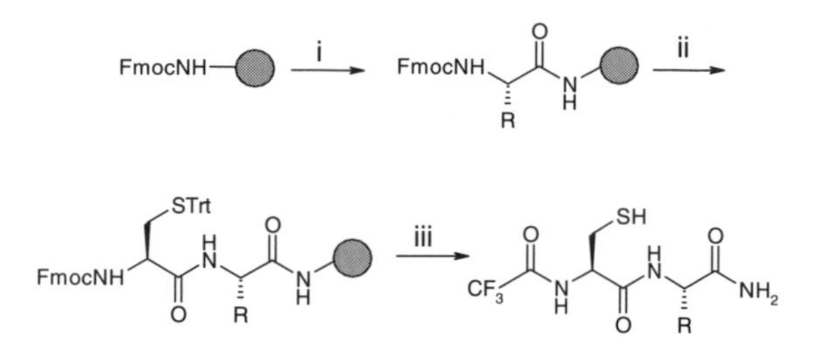

Reagents:
i) Piperidine (20% in DMF); N-Fmoc-aminoacid; HOBt, HBTU, NMP;
ii) Piperidine(20% in DMF); N-Fmoc-S-trityl cysteine; HOBt, HBTU, NMP;
iii) Piperidine(20% in DMF); TFAA, TEA, King Reagent

Scheme 9.2 Solid phase synthesis of cysteine-derived MMPIs.

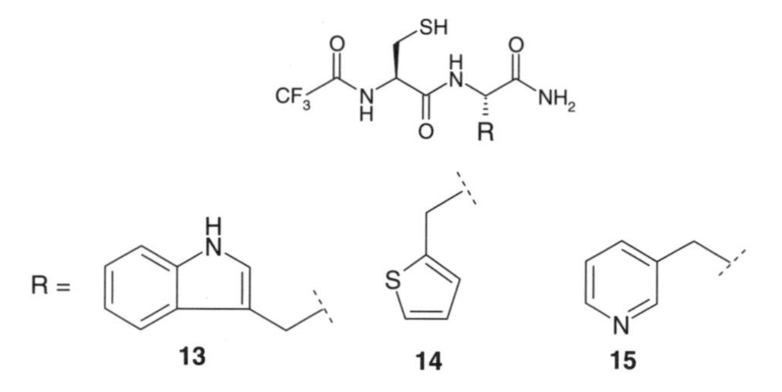

Figure 9.11 Collagenase-selective MMPIs discovered by Foley *et al.* (1996). Collagenase IC_{50} for **13** = 5 μM, for **14** = 4 μM, and for **15** = 4 μM.

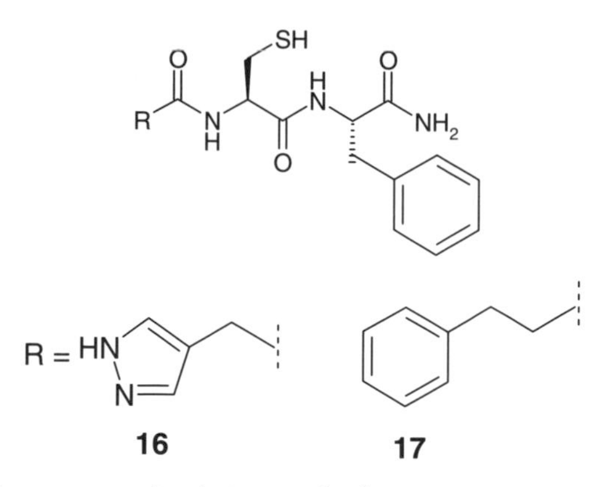

Figure 9.12 Collagenase- and gelatinase-selective MMPIs discovered by Foley *et al.* (1996). IC_{50}s for **16**: gelatinase = 71 nM, collagenase =26 nM; and IC_{50}s for **17**: gelatinase = 38 nM, collagenase = 3498 nM.

groups. Collagenase selectivity was achieved where trifluoroacetyl groups or groups derived from heteroaromatic acids were present (**12**, **16**), and gelatinase selectivity was observed in cases where large lipophilic groups were present at the N terminus (e.g., **17**).

Workers at British Biotechnology have investigated solid phase combinatorial synthesis of MMPIs in a comprehensive fashion. In one case, a novel linker (Scheme 9.3) was constructed (Floyd, Lewis and Whittaker, 1996) to act as a template for the synthesis of hydroxamic acid libraries which possess peptidic and peptidomimetic moieties (Floyd *et al.*, 1996a,b). The synthesis of two kinds of library was reported using this hydroxylamine linker. The first of these libraries, containing 500 tripeptides, was prepared using the split-mix method (Scheme 9.4). Ten mixtures were prepared, each containing 50 compounds, and screened against collagenase, stomelysin and gelatinase, though none of the constituents of the active mixtures displayed greater potency than the existing lead (**18**, Figure 9.13). The second library synthesized was composed of sulfonamide hydroxamic acids (Scheme 5). Several compound arrays were prepared and a hit, **19**, was identified, which displayed activity against stromelysin (Figure 9.14).

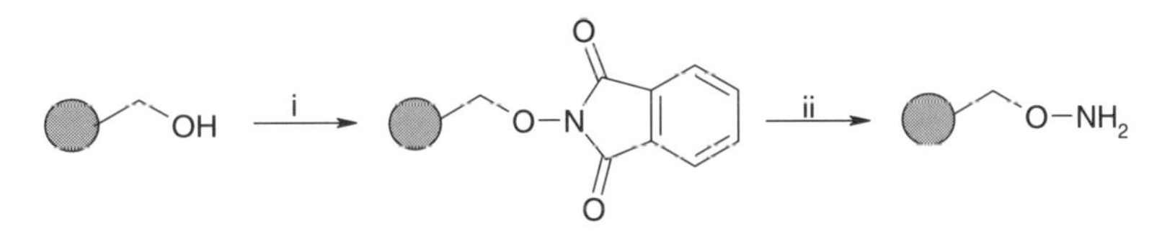

Reagents:
i) THF, PPh_3, DEAD, *N*-hydroxyphthalimide, RT.
ii) Ethanol/THF, hydrazine, RT

Scheme 9.3 Preparation of a novel hydroxylamine linker for solid phase hydroxamic acid synthesis.

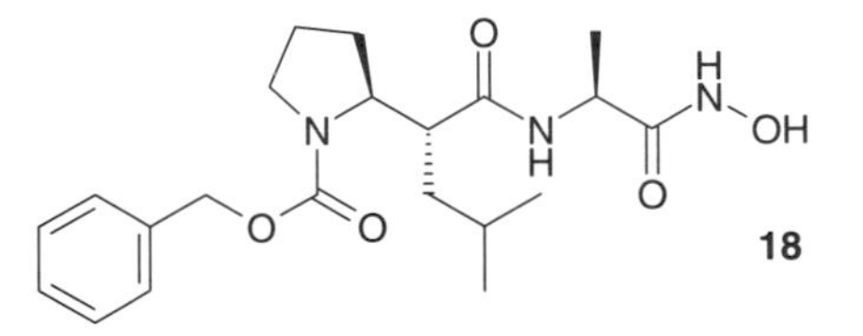

Figure 9.13 BB-3505. IC_{50}s: MMP-1 = 8000 nM, MMP-3 = 3500 nM, and MMP-2 = 8000 nM.

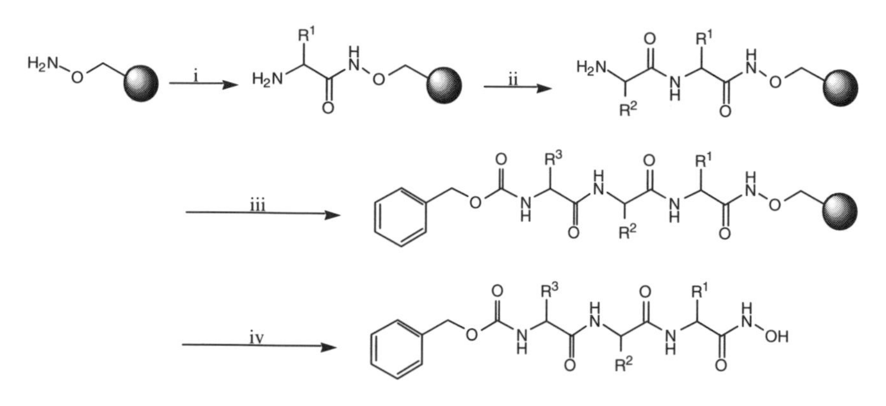

Reagents:
i) Fmoc–AA_1 (x 10), HOBt, DIC, DMF, RT; Piperidine (30% in DMF).
ii) Fmoc–AA_2 (x 10), HOBt, DIC, DMF, RT; Piperidine (30% in DMF).
iii) Cbz–AA_3 (x 5), HOBt, DIC, DMF.
iv) Phenol (5%), TIS (2%), H_2O (5%) in TFA

Scheme 9.4 Solid phase synthesis of hydroxamic acid MMPIs.

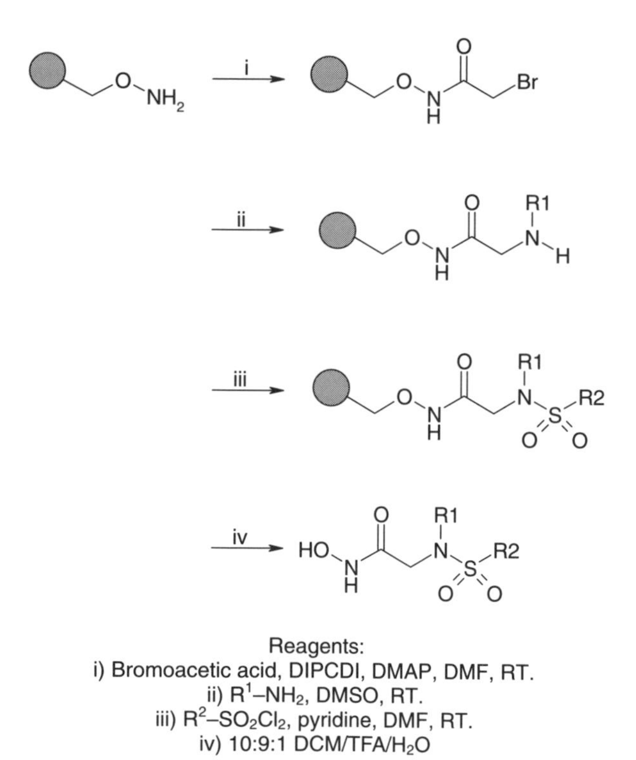

Scheme 9.5 Preparation of sulfonamide hydroxamic acids using solid phase synthesis.

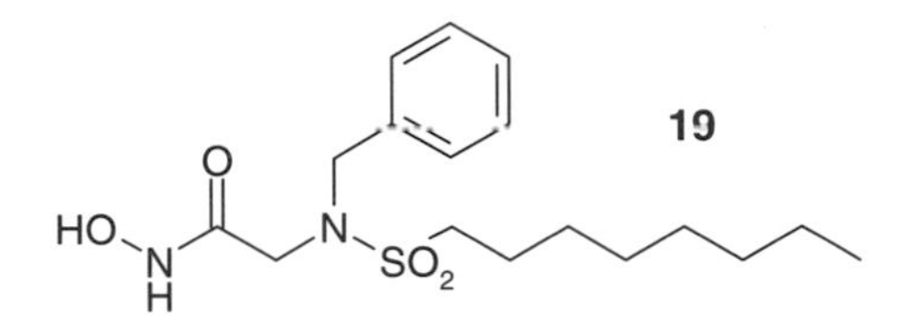

Figure 9.14 BB-3345 identified by combinatorial synthesis. IC_{50} for stromelysin-1 = 200 nM.

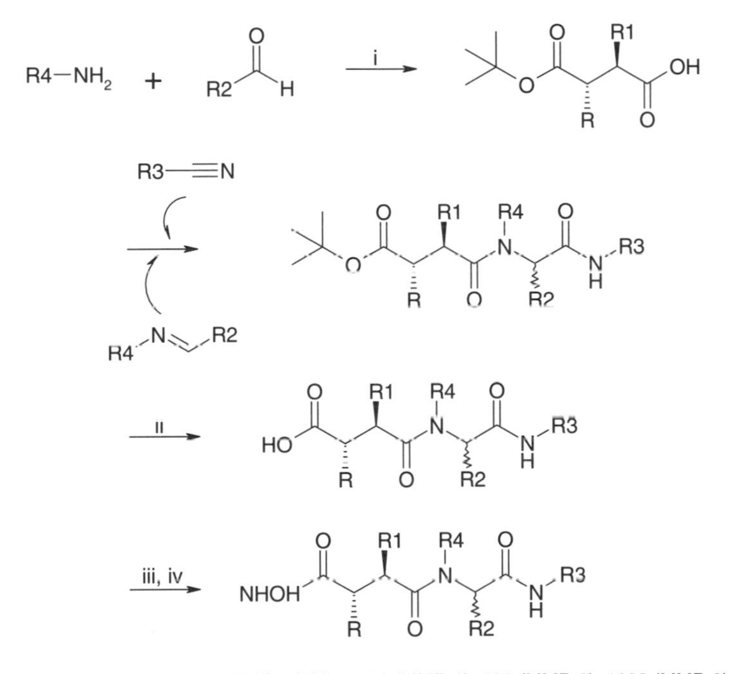

20: R = n-hexyl, R3 = t-butyl; IC_{50} (nM) = 500 (MMP-1), 200 (MMP-2), 1000 (MMP-3)
21: R = n-heptyl, R3 = t-butyl; IC_{50} (nM) = 700 (MMP-1), 50 (MMP-2), 1000 (MMP-3)
22: R = n-octyl, R3 = methyl; IC_{50} (nM) = 100 (MMP-1), 0.7 (MMP-2), 100 (MMP-3)

Reagents:
i) MeOH, rt.
ii) TFA/DCM, 4 °C.
iii) EDAC.HCl, HOBt, H_2NOCH_2Ph; DMF, 0°C.
iv) H_2, Pd-C, EtOH, rt.

Scheme 9.6 Ugi 4-component coupling reaction applied to solution phase synthesis of MMPIs.

British Biotechnology researchers have investigated also solution phase combinatorial synthesis of MMPIs. In this case the Ugi four-component coupling reaction (Short, Ching and Mjalli, 1997), a subject of extensive literature coverage recently, was employed (Miller, 1996). Condensation of carboxylic acids, amines (including ammonia), aldehydes and isocyanides furnished libraries of peptidomimetics in moderate yields (Scheme 9.6) which displayed activity against collagenase, gelatinase and stromelysin (e.g. **20**, **21**, and **22**).

Solid phase synthesis of hydroxamate derivatives (Scheme 9.7) has been the subject of a recent communication by Chen and Spatola (1997). Synthesis of compounds based upon **23** using *p*-methylbenzylhydrylamine (p-MBHA) resin was accomplished; the product hydroxamates released from the resin were being evaluated for MMP activity at the time of writing.

Affymax have investigated solid phase synthesis of phosphinylpeptides which display MMP activity (Campbell, 1996). The synthetic route to these compounds is similar to that previously employed by Affymax for the synthesis of thermolysin-inhibitory peptidylphophonates, as shown in Scheme 9.8 (Campbell *et al.*, 1995). Activities of these compounds against the MMPs have not been published. Similar work is under investigation at Versicor (Patel, 1996).

Finally, researchers at DuPont Merck recently published their investigations into the combinatorial synthesis of MMPIs on solid phase using structure-based ligand design (Rockwell *et al.*, 1996). A three-step synthesis of *N*-carboxyalkyl amino acid derivatives was devised involving anchoring to the polymer support, reductive amination, and amidation (Scheme 9.9). In all cases the products were isolated as 1:1 epimeric mixtures. Previous studies had shown that high potency against MMP-3 is observed were R^1 is methyl and R^2 is 2-phenethyl. These groups were retained for the combinatorial investigations, and a series of diverse amines was incorporated into the product template (position R^3). Over 100 derivatives were screened, with weak activity against

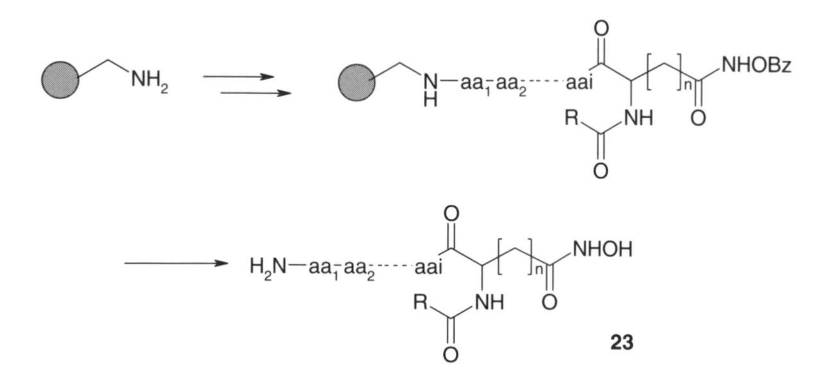

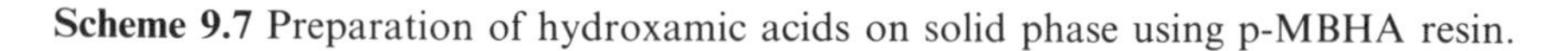
Scheme 9.7 Preparation of hydroxamic acids on solid phase using p-MBHA resin.

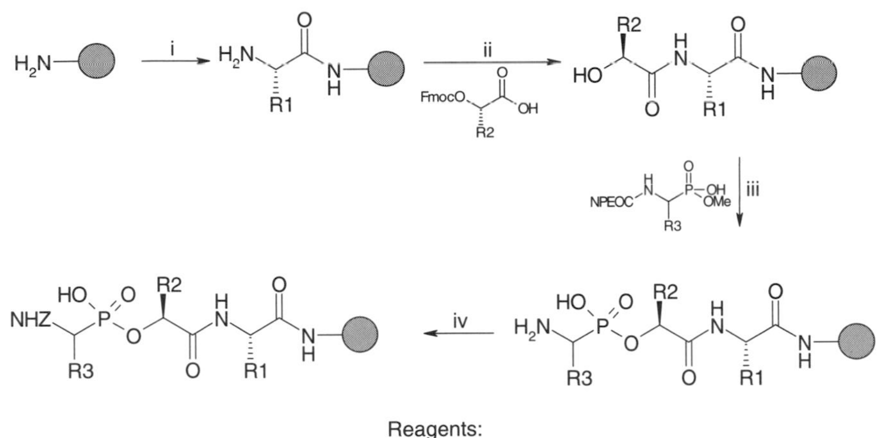

Reagents:
i) HBTU, HOBt, DIPEA, NMP (x 1); PyProP, DIPEA, NMP (x 1); Piperidine/NMP (30:70)
ii) HBTU, HOBt, DIPEA, NMP (x 1); PyProP, DIPEA, NMP (x 1); Piperidine/NMP (30:70)
iii) tris(4-chlorophenyl)phosphine, DIAD, DIPEA, THF; DBU (5%) in NMP
iv) CbzCl, DIPEA, dioxane; thiophenol/TEA/dioxane (1:2:2); TES, TFA

Scheme 9.8 Solid phase synthesis of phosphinylpeptides.

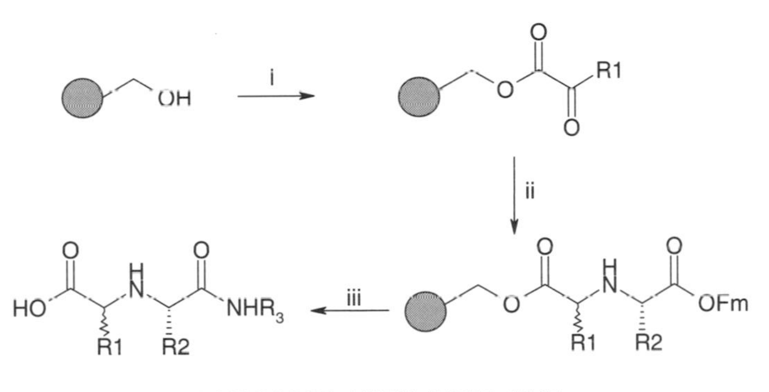

i. HOOCCOR, HBTU, DIPEA, DMF.
II. HCl $NH_2CH(R^2)COOFm$, $NaCNBH_3$, DIPEA, HOAc, DMF.
iii. Piperidine (20%)/DMF; HSpfp, DIC/DMF:DCM; NH_2R^3, DIPEA, HOBt, DMF;
HCl (4N), Dioxane

Scheme 9.9 Synthesis of *N*-carboxyalkyl amino acids on solid phase.

MMP-3 being observed for compound **24** (10% inhibition at 100 μM). Stereospecific inhibition was observed: the individual epimers of **24** were separated and displayed different potencies in the enzyme assays. At a concentration of 100 μM, **25** inhibited MMP-3 by 6%, whereas **26** displayed 33% inhibition (Figure 9.15). Subsequent investigations let to the identification of **27** (Figure 9.16), which showed good potency against MMP-3 (148 nM) and MMP-8 (1.9 nM).

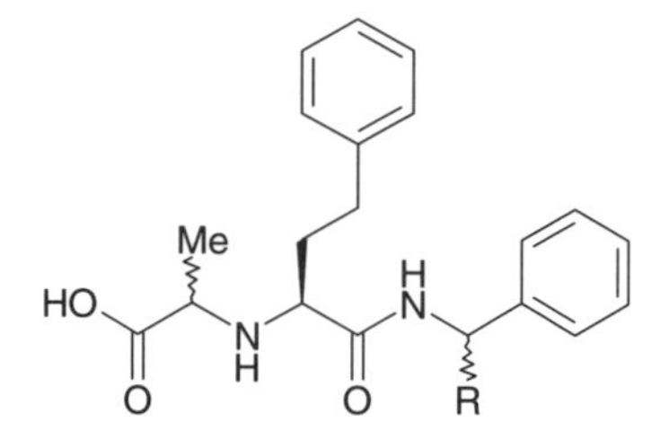

Figure 9.15 Weak MMPIs identified by Rockwell *et al.* (1996).

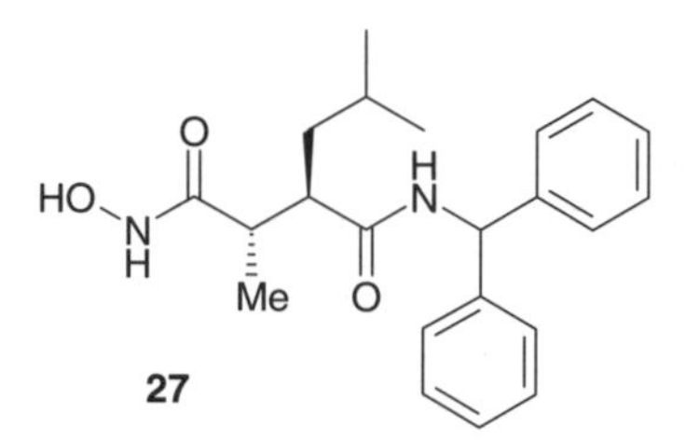

Figure 9.16 Selective inhibitor of MMP-8 identified by DuPont Merck (Rockwell *et al.*, 1996).

9.4 CONCLUSIONS

The work described above illustrates the prominence of combinatorial chemistry in an exciting and increasingly important area of drug discovery. It is evident from the examples that the use of solid and solution phase combinatorial chemistry has been successful in helping several research groups identify inhibitors of matrix metalloproteinases. It is hoped that the area will continue to blossom in the new millenium.

ACKNOWLEDGEMENTS

This chapter refers to material to which the author had access as an employee of Chiroscience Limited, Cambridge Science Park, Milton Road, Cambridge, CB4 4WE, England. All this material has previously been published at conferences.

REFERENCES

Adinolfi, M. Barone, G., De Napoli, L., Iadonisi, A. and Piccialli, G. (1996) Solid phase synthesis of oligosaccharides, *Tetrahedron Letters*, **37**, 5007–10.

Allin S.M. and Shuttleworth, S.J. (1996) The preparation and first application of a polymer-supported "Evans" oxazolidinone, *Tetrahedron Letters*, **37**, 8023–6.

Anderson, R.C., Jarema, M.A., Shapiro, M.J., Stokes, J.P. and Ziliox, M. (1996) NMR techniques in combinatorial chemistry, *Bruker Reviews*, 39–41.

Bailey, N., Dean, A.W., Judd, D.B., Middlemiss, D., Storer, R.C. and Watson, S.P. (1996) A convenient procedure for the solution phase preparation of 2-aminothiazole combinatorial libraries, *Bioorganic and Medicinal Chemistry Letters*, **6**, 1409–14.

Baldwin, J.J., Burbaum, J.J., Henderson, I. and Ohlmeyer, M.H.J. (1995) Synthesis of a small molecule combinatorial library encoded with molecular tags, *Journal of the American Chemical Society*, **117**, 5588–9.

Balkenhohl, F., von dem Bussche–Hünnefeld, C., Lansky, A. and Zechel, C. (1996) Combinatorial synthesis of small molecules, *Angewandte Chemie International Edition in English*, **35**, 2288–337.

Basu and Wickstrom, E. (1995) Solid phase synthesis of a δ-peptide-phosphoroithioate oligodeoxynucleotide conjugate from two arms of a polyethylene glycol–polystyrene support, *Tetrahedron Letters*, **36**, 4943–6.

Baxter, A.D., Bird, J., Bhogal, R., Massil, T., Minton, K.J., Montana, J. and Owen, D.A. (1997) A novel series of matrix metalloproteinase inhibitors for the treatment of inflammatory disorders, *Bioorganic and Medicinal Chemistry Letters*, **7**, 897–902.

Beckett, R.P., Bellamy, C.L., Floyd, C.D., Harnett, L.A., Martin, F., Patel, S., Saroglou, L., Thompson, A.J. and Whittaker, M. (1996) Matrix metalloproteinase inhibitors: examination of the S1′ pocket utilising the Ugi four component condensation, presented at the RSC Bioorganic Subject Group Meeting, Crieff, UK, September.

Beckett, R.P., Davidson, A.H., Drummond, A.H., Huxley, P. and Whittaker, M. (1996) Recent advances in matrix metalloproteinase inhibitor research, *Drug Discovery Today*, **1**, 16–20.

Beck-Sickinger, A.G., Dürr, H. and Jung, G. (1991) Semiautomated T–bag peptide synthesis using 9–fluoromethoxycarbonyl strategy and benzotriazol–1–yl–tetramethyluronium tetrafluoroborate activation, *Peptide Research*, **4**, 88–95.

Bird, J., Harper, G.P., Hughes, I., Hunter, D.J., Karran, E.H., Markwell, R.E., Miles–Williams, A.J., Rahman, S.S. and Ward, R.W. (1995) Inhibitors of human collagenase: dipeptide mimetics with lactam and azalactam moieties at the P_2'/P_3' position *Bioorganic and Medicinal Chemistry Letters*, **5**, 2593–8.

Boger, D.L., Chai, W., Ozer, R.S. and Anderson, C. (1997) Solution phase combinatorial synthesis via the olefin metathesis reaction, *Bioorganic and Medicinal Chemistry Letters*, **7**, 463–8.

Boger, D.L., Tarby, C.M., Myers, P.L., and Caporale, L.H. (1996) Generalised dipeptidomimetic template: solution phase parallel synthesis of combinatorial libraries, *Journal of the American Chemical Society*, **118**, 2109–10.

Boojamra, C.G., Burow, K.M. and Ellman, J.A. (1995) An expedient and high-yielding method for the solid phase synthesis of diverse 1,4–benzodiazepine–2,5–diones, *Journal of Organic Chemistry*, **60**, 5742–3.

Boons, G.J., Heskamp, B. and Hout, F. (1996) Vinyl glycosides in oligosaccharide synthesis: a strategy for the preparation of trisaccharide libraries based on latent–active glycosylation, *Angewandte Chemie International Edition in English*, **35**, 2845–7.

Borman, S. (1997) Combinatorial chemistry, *Chemical and Engineering News*, 43–62.

Broadhurst, M. J. (1993) European Patent Application 1993 EP–497, 192-A.

Brown, A.R., Rees, D.C., Rankovic, Z. and Morphy, J.R. (1997) Synthesis of tertiary amines using a polystyrene (REM) resin, *Journal of the American Chemical Society*, **119**, 3288–95.

Brown, F.K., Brown, P.J. Bickett, D.M., Chambers, C.L., Deaton, D.N., Drewry, D., Foley, M., McElroy, A.B., Gregson, M., McGeehan, G.M., Myers, P.L., Norton, D., Salovich, J.M. Schoenen, F.J. and Ward, P. (1994) Matrix metalloproteinase inhibitors containing a (carboxyalkyl)amino zinc ligand: modification of the P_1 and P_2' residues, *Journal of Medicinal Chemistry*, **37**, 674–88, and references therein.

Bunin, B.A. and Ellman, J.A. (1992) A general and expedient method for the solid phase synthesis of 1,4–benzodiazepine derivatives, *Journal of the American Chemical Society*, **114**, 10997–8.

Bunin, B.A., Plunkett, M.J. and Ellman, J.A. (1994) The combinatorial synthesis and chemical and biological evaluation of a 1,4–benzodiazepine library, *Proceedings of the National Academy of Sciences of the USA*, **91**, 4708–12.

Burgess, K. and Lim, D. (1997) Resin type can have imported effects on solid phase asymmetric alkylation reactions, *Journal of the Chemical Society: Chemical Communications*, 785–6.

Caldwell, C.G., Sahoo, S.P., Eversole, R.R., Lanza, T.J., Mills, S.G., Niedwiecki, L. Izquierdo–Martin, M., Chang, B.C., Harrison, R.K., Kuo, D.W., Lin, T.-Y., Stein, R.L., Durette, P.L., and Hagmann, W.K. (1996) Phosphinic acid inhibitors of matrix metalloproteinases, *Bioorganic and Medicinal Chemistry Letters*, **6**, 323–6.

Campbell, D.A. (1996) Application of combinatorial strategies towards metalloprotein-ase inhibition identification, presented at the Combinatorial Chemistry Symposium 2, University of Exeter, UK, July.

Campbell, D.A., Bermak, J.C., Burkoth, T.S. and Patel, D.V. (1995) A transition state analogue inhibitor combinatorial library, *Journal of the American Chemical Society*, **117**, 5381–2.

Carell, T., Whitner, E.A., Sutherland, A.J., Rebek, J., Dunayevskiy, Y.M. and Vouros, P. (1995) New promise in combinatorial chemistry: synthesis, characterization, and screening of small-molecule libraries in solution, *Chemistry & Biology*, **2**, 171–83.

Castelhano, A.L., Billdeau, R., Dewdney, N., Donnelly, S., Horne, S., Kurz, L.J., Liak, T.J., Marton, R., Uppington, R., Yuan, Z. and Krantz, A. (1995) Novel indolactam based inhibitors of matrix metalloproteinases, *Bioorganic and medicinal chemistry letters*, **13**, 1415–8.

Chan, T.Y., Chen, R., Sofia, M.J., Smith, B. and Glennon, D. (1997) High throughput on-bead monitoring of solid phase reactions by diffuse reflectance infrared Fourier transform spectroscopy, *Tetrahedron Letters*, **38**, 2821–4.

Chapman, K.T., Kopka, I.E., Durette, P.L., Esser, C.K., Lanza, T.J., Izquierdo–Martin, M., Niedwiecki, L., Chang, B., Harrison, R.K., Kuo, D.W., Lin, T.-Y., Stein, R. and Hagmann, W.K. (1993) Inhibition of matrix mettaloproteinases by *N*-carboxyalkyl peptides, *Journal of Medicinal Chemistry*, **36**, 4293.

Chen, C., Randall, L.A.A., Miller, R.B., Jones, A.D. and Kurth, M.J. (1994) "Analogous" organic synthesis of small-compound libraries: validation of combinatorial chemistry in small-molecule synthesis, *Journal of the American Chemical Society*, **116**, 2661–2.

Chen, J.J. and Spatola, A.F. (1997) Solid phase synthesis of peptide hydroxamic acids, *Tetrahedron Letters*, **38**, 1511–4.

Cheng, S., Comer, D.D., Williams, J.P., Myers, P.L. and Boger, D.L. (1996) Novel solution phase strategy for the synthesis of chemical libraries containing small organic molecules, *Journal of the American Chemical Society*, **118**, 2567–73.

Cohen, N.C., Blaney, J.M., Humblet, C., Gund, P. and Barry, D.C. (1990) Molecular modeling software and methods for medicinal chemistry, *Journal of Medicinal Chemistry*, **33**, 883–94.

Cramer, R.D., Clark, R.D., Patterson, D.E. and Ferguson, A.M. (1996) Bioisosterism as a molecular diversity descriptor: steric fields of single "topomeric" conformers, *Journal of Medicinal Chemistry*, **39**, 3060–8.

Davidson, A.H., Drummond, A.H., Galloway, W.A. and Whittaker, M. (1997) The inhibition of matrix metalloproteinase enzymes, *Chemistry & Industry*, 258–61.

Davies, J.S. (1992) in *Solid Supports and Catalysis in Organic Synthesis* (ed. K. Smith), Ellis Horwood, Chichester, p. 195.

Deshpande, M.S. (1994) Formation of carbon–carbon bond on solid support: application of the stille reaction, *Tetrahedron Letters*, **35**, 5613–6.

De Witt, S.H., Kiely, J.S., Stankovic, C.J., Schroeder, M.C., Cody, D.M.R. and Pavia, M.R. (1993) "Diversomers": an approach to nonpeptide, nonoligomeric chemical diversity, *Proceedings of the National Academy of Sciences of the USA*, **90**, 6909–13.

Dickens, J.P., Crimmin, M.J. and Beckett, R.P. (1994) PCT Patent Application, WO9402447.

Doyle, P.M. (1995) Combinatorial chemistry in the discovery and development of drugs, *Journal of Chemical Technology and Biotechnology*, **64**, 317–24.

Egner, B.J., Cardno, M. and Bradley, M. (1995) Linkers for combinatorial chemistry and reaction analysis using solid phase *in situ* mass spectrometry, *Journal of The Chemical Society: Chemical Communications*, 2163–4.

Egner, B.J., Langley, G.J. and Bradley, M. (1995) Solid phase chemistry: direct monitoring by matrix assisted laser desorption/ionization time of flight mass spectrometry. A tool for combinatorial chemistry, *Journal of Organic Chemistry*, **60**, 2652–3.

Eichler, J. and Houghten, R.A. (1995) Generation and utilization of synthetic combinatorial libraries, *Molecular Medicine Today*, **1**, 174–80.

Emery, F., Bisang, C., Favre, M., Jiang, L. and Robinson, J.A. (1996) A template for the solid-phase synthesis of conformationally restricted protein loop mimetics, *Journal of The Chemical Society: Chemical Communications*, 2155–6.

Esser, C.K., Kopka, I.E., Durette, P.L., Harrison, R.K., Niedwiecki, L., Izquierdo–Martin, M., Stein, R.L. and Hagmann, W.K. (1995) Inhibition of matrix metalloproteinases by *N*-carboxyalkyl peptides containing extended alkyl residues at P_1', *Bioorganic and Medicinal Chemistry Letters*, **5**, 539–42.

Ferguson, A.M., Patterson, D.E., Garr, C.D. and Underliner, T.L. (1996) Designing chemical libraries for lead discovery, *Journal of Biomolecular Screening*, **1**, 65–73.

Fitch, W.L., Detre, G. and Holmes, C.P. (1994) High resolution ^{1}H NMR in solid phase organic synthesis, *Journal of Organic Chemistry*, **59**, 7955–6.

Floyd, C.D., Harnett, L.A., Lewis, C.N., Patel, S.R., Whittaker, M. and Floyd, C.D. (1996b) Solid phase synthesis of hydroxamic acid inhibitors of matrix metalloproteinases, presented at Combinatorial Chemistry Symposium 2 University of Exeter, UK, July.

Floyd, C.D., Lewis, C.N., Patel, S.R., and Whittaker, M. (1996a) A method for the synthesis of hydroxamic acids on solid phase, *Tetrahedron Letters*, **37**, 8045–8.

Floyd, C.D., Lewis, C.N., Whittaker, M. (1996) More leads in the haystack, *Chemistry in Britain*, 31–5.

Foley, M.A., Hassman, A.S., Drewry, D.H., Greer, D.G., Warner, C.D., Feldman, P.L., Bermann, J., Bickett, D.M., McGeehan, G.M., Lambert, M.H. and Green, M. (1996) Rapid synthesis of novel dipeptide inhibitors of human collagenase and gelatinase using solid phase chemistry, *Bioorganic and Medicinal Chemistry Letters*, **6**, 1905–10.

Früchtel, K.S. and Jung, G. (1996) Organic chemistry on solid supports, *Angewandte Chemie International Edition in English,* **35**, 17–42.

Furka, A., Sebestyen, F., Asgedom, M. and Dibo, G. (1991) General method for rapid synthesis of multicomponent peptide mixtures, *International Journal of Peptide and Protein Research,* **37**, 487–93.

Gallop, M.A., Barrett, R.W., Dower, W.J., Fodor, S.P.A. and Gordon, E.M. (1994) Applications of combinatorial technologies to drug discovery 2. Combinatorial organic synthesis, library screening strategies and future directions, *Journal of Medicinal Chemistry*, **37**, 1385–401.

Gayo, L.M. and Suto, M.J. (1997a) Traceless linker: oxidative activation and displacement of a sulfur-based linker, *Tetrahedron Letters*, **38**, 211–14.

Gayo, L.M. and Suto, M.J. (1997b) Ion-exchange resins for solution phase parallel synthesis of chemical libraries, *Tetrahedron Letters*, **38**, 513–16.

Geysen, H.M., Meloen, R.H. and Barteling, S. (1984) Use of peptide synthesis to prove viral antigens for epitopes to a resolution of a single amino acid, *Proceedings of the National Academy of Sciences of the USA*, **81**, 3988–4002.

Geysen, H.M., Rodda, S.J. and Mason, T. (1986) *A priori* delineation of a peptide which mimics a discontinuous antigenic determinant, *Molecular Immunology*, **23**, 709–15.

Geysen, H.M., Rodda, S.J., Mason, T., Tribbick, G. and Schoofs, P. (1987) Strategies for epitope analysis using a peptide synthesis, *Journal of Immunology Methods*, **102**, 259–61.

Goff, D.A. and Zuckerman, R.N. (1995) Solid phase synthesis of defined 1,4–benzodiazepine–2,5–dione mixtures, *Journal of Organic Chemistry*, **60**, 5744–5.

Gordon, E.M., Barrett, R.W., Dower, W.J., Fodor, S.P.A. and Gallop, M.A. (1994) Applications of combinatorial technologies to drug discovery 1. Background and peptide combinatorial libraries, *Journal of Medicinal Chemistry*, **37**, 1233–51.

Goulet, J.L., Kinneary, J.F., Durette, P.L., Stein, R.L., Harrison, R.K., Izquierdo–Martin, M., Kuo, D.W., Lin, T.-Y. and Hagmann, W.K. (1994) Inhibition of stromelysin-1 (MMP-3) by peptidyl phosphinic acids, *Bioorganic and Medicinal Chemistry Letters*, **4**, 1221–4.

Gowravaram, M.R., Tomczuk, B.E., Johnson, J.S., Delecki, D., Cook, E.R., Ghose, A.K., Mathiowetz, A.M., Spurlino, J.C., Rubin, B., Smith, D.L., Pulvino, T., and Wahl, R.C. (1995) Inhibition of matrix metalloproteinases by hydroxamates containing heteroatom-based modifications of the P1′ group, *Journal of Medicinal Chemistry*, **38**, 2570–81.

Guiles, J.W., Johnson, S.G. and Murray, W.V. (1996) Solid phase Suzuki coupling for C–C bond formation, *Journal of Organic Chemistry*, **61**, 5169.

Hamper, B.C. Dukesherer, D.R. and South, M.S. (1996) Solid phase syntheses of proline analogs *via* a three component 1,3–dipolar cycloaddition, *Tetrahedron Letters*, **37**, 3671–4.

Han, H. and Janda, K.D. (1996) Azatides: solution and liquid phase syntheses of a new peptidomimetic, *Journal of the American Chemical Society*, **118**, 2539–44.

Han, Y., Walker, S.A., and Young, R. (1996) Silicon directed *ipso*-substitution of polymer bound arylsilanes: preparation of biaryls via the Suzuki cross-coupling reaction, *Tetrahedron Letters*, **37**, 2703–6.

Haskins, N.J., Hunter, D.J., Organ, A.J., Rahman, S.S. and Thom, C. (1995) Combinatorial chemistry: direct analysis of bead surface associated materials, *Rapid Communications in Mass Spectrometry*, **9**, 1437–9.

Hermkens, P.H.H., Ottenheijm, H.C.J. and Rees, D.C. (1996) Solid phase organic reactions: a review of the recent literature, *Tetrahedron*, **52**, 4527–54.

Hermkens, P.H.H., Ottenheijm, H.C.J. and Rees, D.C. (1997) Solid phase organic reactions II: a review of the literature Nov 95–Nov 96, *Tetrahedron*, **53**, 5643–78.

Hiroshige, M., Hauske, J.R. and Zhou, P. (1995) Formation of C–C bond in solid phase synthesis using a Heck reaction, *Tetrahedron Letters*, **36**, 4567–70.

Hoekstra, W.J., Maryanoff, B.E., Andrade–Gordon, P., Cohen, J.H., Constanzo, M.J., Damiano, B.P., Haer, B.J., Harris, B.D., Kauffman, J.A., Keane, P.M., McComsey, D.F., Villiani, F.J. and Yabut, S.C. (1996) Solid–phase parallel synthesis applied to lead optimization: discovery of potent analogues of the GPIIb/IIIa antagonist RWJ–50042, *Bioorganic and Medicinal Chemistry Letters*, **6**, 2371–6.

Holmes, C.P., Chinn, J.P., Look, G.C., Gordon, E.M. and Gallop, M.A. (1995) Strategies for combinatorial organic synthesis: solution and polymer-supported synthesis of 4-thiazolidinones and 4-metathiazanones derived from amino acids. *Journal of Organic Chemistry*, **60**, 7328–33.

Houghten, R.A. (1985) General method for the solid phase synthesis of large numbers of peptides; specificity of antigen–antibody interaction at the level of individual amino acids, *Proceedings of the National Academy of Sciences of the USA*, **82**, 5131–5.

Hughes, I. (1996) Application of polymer-bound phosphonium salts as traceless supports for solid phase synthesis, *Tetrahedron Letters*, **37**, 7595–8.

Hughes, I., Harper, G.P., Karran, E.H., Markwell, R.E. and Miles–Williams, A.J. (1995) Synthesis of thiphenol derivatives as inhibitors of human collagenase, *Bioorganic and Medicinal Chemistry Letters*, **5**, 3039–42.

Hunt, J.A. and Roush, W.R. (1996) Solid phase synthesis of 6–deoxyoligosaccharides, *Journal of the American Chemical Society*, **118**, 9998–9.

Kaldor, S.W., Siegel, M.G., Fritz, J.E., Dressman, B.A. and Hahn, P.J. (1996) Use of solid supported nucleophiles and electrophiles for the purification of non-peptide small molecule libraries, *Tetrahedron Letters*, **37**, 7193–6.

Koh, J.S. and Ellman, J.A. (1996) Palladium–mediated three-component coupling strategy for the solid phase synthesis of tropane derivatives, *Journal of Organic Chemistry*, **61**, 4494–5.

Konings, D.A.M., Wyatt, J.R., Ecker, D.J. and Freier, S.M. (1996) Deconvolution of combinatorial libraries for drug discovery: theoretical comparison of pooling strategies, *Journal of Medicinal Chemistry*, **39**, 2710–9.

Kumar, P., Bhatia, D., Garg, B.S. and Gupta, K.C. (1994) An improved method for synthesis of biotin phosphoramidites for solid phase biotinylation of oligonucleotides, *Bioorganic and Medicinal Chemistry Letters*, **4**, 1761–76.

Lam, K.S., Lebl, M. and Krchnak, V. (1997) The "one-bead-one-compound" Combinatorial Library Method, *Chemical Reviews*, **97**, 411–48.

Le Bec, C. and Wickstrom, E. (1996) Stereospecific Grignard activated solid phase synthesis of DNA methylphosphonate dimers, *Journal of Organic Chemistry*, **61**, 510–18.

Levy, D.E., Lapierre, F., Liang, W., Ye, W., Lange, C.W., Li, X., Grobelny, D., Casabonne, M., Tyrrell, D., Holme, K., Nadzan, A. and Galardy, R.E. (1998) Matrix metalloproteinase inhibitors: a structure activity study, *Journal of Medicinal Chemistry*, **41**, 199–223.

Lin, M. and Shapiro, M.J. (1996) Mixture analysis in combinatorial chemistry: application of diffusion–resolved NMR spectroscopy, *Journal of Organic Chemistry*, **61**, 7617–9.

Look, G.C., Holmes, C.P., Chinn, J.P. and Gallop, M.A. (1994) Methods for combinatorial organic synthesis: the use of fast ^{13}C NMR analysis for gel phase reaction monitoring, *Journal of Organic Chemistry*, **59**, 7588–90.

McGuigan, C., Turner, S., Mahmood, N. and Hay, A.J. (1996) A rapid synthesis of some 5′–amino nucleosides and nucleotides as potential antiviral compounds, *Bioorganic and Medicinal Chemistry Letters*, **6**, 2445–8.

Maud, J.M. (1992) *Solid Supports and Catalysis in Organic Synthesis* (ed. K. Smith), Ellis Horwood, Chichester, p. 40, and references therein.

Merrifield, B. (1986) Solid phase synthesis, *Science*, 232, 341–7.

Merrit, A.T. (1995) presented at Combinatorial Chemistry Symposium 1, University of Exeter, UK, July.

Miller, A. (1996) presented at the RSC Bioorganic Subject Group Meeting, Crieff, UK, September.

Miller, A., Askew, M., Beckett, R.P., Bellamy, C.L., Bone, E.A., Coates, R.E., Davidson, A.H., Drummond, A.H., Huxley, P., Martin, F.M., Saroglou, L., Thompson, A.J., van Dijk, S.E. and Whittaker, M. (1997) Inhibition of matrix metalloproteinases: an examination of the S_1 pocket, *Bioorganic and Medicinal Chemistry Letters*, **7**, 193–8.

Mjalli, A.M.M., Sarshar, S. and Baiga, T. (1996) Solid phase synthesis of pyrroles derived from a four component condensation, *Tetrahedron Letters*, **37**, 2943–6.

Moon, H., Schore, N.E. and Kurth, M.J. (1992) A polymer-supported chiral auxiliary applied to the iodolactonization reaction: preparation of γ-butyrolactones, *Journal of Organic Chemistry*, **57**, 6088–9.

Moon, H., Schore, N.E. and Kurth, M.J. (1994) A polymer-supported C_2-symmetric chiral auxiliary: preparation of non-racemic 3,5-disubstituted γ-Butyrolactones, *Tetrahedron Letters*, **35**, 8915–8.

Morphy, J.R., Beeley, N.R.A., Boyce, B.A., Leonard, J., Mason, B., Millican, A., Millar, K., O'Connell, J.P. and Porter, J.R. (1994) Potent and selective inhibitors of gelatinase-A.2. Carboxylic and phosphinic acid derivatives, *Bioorganic and Medicinal Chemistry Letters*, **4**, 2747–52.

Nash, I.A., Bycroft, B.W. and Chan, W.C. (1996) Dde – a selective primary amine protecting group: a facile solid phase synthetic approach to polyamine conjugates, *Tetrahedron Letters*, **37**, 2625–8.

Nefzi, A., Ostresh, J.M. and Houghten, R.A. (1997) The current status of heterocyclic combinatorial libraries, *Chemical Reviews*, **97**, 449–72.

Nestler, H.P., Bartlett, P.A. and Still, W.C. (1994) A general method for the molecular tagging of encoded combinatorial chemistry libraries, *Journal of Organic Chemistry*, **59**, 4723–4.

Neuner, P. (1996) New non-nucleosidic phosphoramidite reagent for solid phase synthesis of biotinylated oligonucleotides, *Bioorganic and Medicinal Chemistry Letters*, **6**, 147–52.

Nicolaou, K.C., Wissinger, N., Pastor, J. and DeRoose, F. (1997) A general and highly efficient solid phase synthesis of oligosaccharides. Total synthesis of a heptasaccharide phytoalexin elicitor (HPE), *Journal of the American Chemical Society*, **119**, 449–50.

Nicolaou, K.C., Xiao, X.-Y., Parandoosh, Z., Senyei, A. and Nova, M. (1995) Radiofrequency encoded combinatorial chemistry, *Angewandte Chemie International Edition in English*, **34**, 2289–91.

Nielsen, J. (1994) Combinatorial chemistry, *Chemistry & Industry*, 902–8.

Nuigel, D.A., Jacobs, K., Decicco, C.P., Nelson, D.J., Copeland, R.A. and Hardman, K.D. (1995) Probing the P_3' pocket of stromelysin and piperazic acid analogues, *Bioorganic and Medicinal Chemistry Letters*, **5**, 3053–6.

O'Donnell, M.J., Zhou, C. and Scott, W. (1996) Solid-phase unnatural peptide-synthesis, *Journal of the American Chemical Society*, **118**, 6070–1.

Parlow, J.J. (1996) The use of anion exchange resins for the synthesis of combinatorial libraries containing aryl and heteroaryl ethers, *Tetrahedron Letters*, **37**, 5257–60.

Parlow, J.J. and Normansell, J.E. (1995) Discovery of a herbicidal lead using polymer-bound activated esters in generating a combinatorial library of amides and esters, *Molecular Diversity*, **1**, 266–9, and references therein.

Patel, D.V. (1996) The next step in combinatorial chemistry: from technology to lead discovery and development, presented at the Meeting on Molecular Diversity and Combinatorial Chemistry, San Diego, USA, October.

Patterson, D.E., Cramer, R.D., Ferguson, A.M., Clark, R.D. and Weinberger, L.E. (1996) Neighbourhood behaviour: a useful concept for validation of "molecular diversity", *Journal of Medicinal Chemistry*, **39**, 3049–59.

Pavia, M.R., Sawyer, T.K. and Moos, W.H. (1993) Generation of molecular diversity, *Bioorganic and Medicinal Chemistry Letters*, **3**, 387–96.

Peterson, J.R. (1996) The Optiverse Chemical Library, presented at the Barnett International Combinatorial Chemistry & Molecular Diversity Meeting, San Diego, March.

Pirrung, M.C. (1997) Spatially addressable combinatorial libraries, *Chemical Reviews*, **97**, 463–88.

Pirrung, M.C. and Chen, J. (1995) Preparation and screening against acetylcholinesterase of a non-peptide "indexed" combinatorial library, *Journal of the American Chemical Society*, **117**, 1240–5.

Plunkett, M.J., and Ellman, J.A. (1995) Solid phase synthesis of structurally diverse 1,4–benzodiazpine derivatives using the Stille coupling reaction, *Journal of the American Chemical Society*, **117**, 3306–7.

Porter, J.R., Beeley, N.R.A., Boyce, B.A., Mason, B., Millican, A., Millar, K., Leonard, J., Morphy, J.R. and O'Connell, J.P. (1994) Potent and selective inhibitors of gelatinase-A.1. Hydroxamic acid derivatives, *Bioorganic and Medicinal Chemistry Letters*, **4**, 2741–46.

Robinson, R.P., Raglan, J.A., Cronin, B.J., Donahue, K.M., Lopresti-Morrow, L.L., Mitchell, P.G., Reeves, LM. and Yocum, S.A. (1996) Inhibitors of MMP-1: an examination of $P_1' C_\alpha$ gem-disubstitution in the succinamide hydroxamate series, *Bioorganic and Medicinal Chemistry Letters*, **6**, 1719–24.

Rockwell, A., Melden, M., Copeland, R.A., Hardman, K., Decicco, C.P. and DeGrado, W.F. (1996) Complimentarity of combinatorial chemistry and structure-based ligand design: application to the discovery of novel inhibitors of matrix metalloproteinases, *Journal of the American Chemical Society*, **118**, 10337–8.

Rotella, D.P. (1996) Solid phase synthesis of olefin and hydroxyethylene peptidomimetics, *Journal of the American Chemical Society*, **118**, 12246–7.

Sarshar, S., Siev, D. and Mjalli, A.M.M. (1996) Imidazole libraries on solid support, *Tetrahedron Letters*, **37**, 835–8.

Schulster, M., Wang, P., Paulson, J.C. and Wong, C.-H. (1994) Solid phase chemical-enzymatic synthesis of glycopeptides and oligosaccharides, *Journal of the American Chemical Society*, **116**, 1135–6.

Shimizu, H., Ito, Y., Kanie, O. and Ogawa, T. (1996) Solid phase synthesis of polylactosamine oligosaccharide, *Bioorganic and Medicinal Chemistry Letters*, **6**, 2841–6.

Short, K.M., Ching, B.W. and Mjalli, A.M.M. (1996) The synthesis of hydantoin–4–imides on solid support, *Tetrahedron Letters*, **27**, 7489–92.

Short, K.M., Ching, B.W. and Mjalli, A.M.M. (1997) Exploitation of the Ugi 4CC reaction: preparation of small molecule combinatorial libraries on solid phase, *Tetrahedron*, **53**, 6653–80.

Shuttleworth, S.J. (1996a) Solution phase combinatorial synthesis of MMPIs, presented at the Advances in Drug Discovery Techniques Meeting, Glasgow, UK, June.

Shuttleworth, S.J. (1996b) Combinatorial synthesis at Chiroscience — design and implementation, presented at Combinatorial Synthesis Symposium 2, University of Exeter, UK, July.

Shuttleworth, S.J., Allin, S.M. and Sharma, P.K. (1997) Functionalised polymers: recent developments and new applications in synthetic organic chemistry, *Synthesis*, 1217–1239.

Siegel, M.G., Hahn, P.J., Dressman, B.A., Fritz, J.E., Grunwell, J.R. and Kaldor, S.W. (1997) Rapid purification of small molecule libraries by ion exchange chromatography, *Tetrahedron Letters*, **38**, 3357–60.

Singh, J., Conzentino, P., Cundy, K., Gainor, J.A., Gilliam, C.L., Gordon, T.D., Johnson, J.A., Morgan, B.A., Schneider, E.D., Wahl, R.C. and Whipple, D.A. (1995) Relationship between structure and bioavailability in a series of hydroxamate based metalloproteinase inhibitors, *Bioorganic and Medicinal Chemistry Letters*, **5**, 337–40.

Smith, P.W., Lai, J.Y.Q., Whittington, A.R., Cox, B., Houston, J.G., Stylli, C.H., Banks, M.N. and Tiller, P.R. (1994) Synthesis and biological evaluation of a library containing 1600 amides/esters. A strategy for rapid compound generation and screening, *Bioorganic and Medicinal Chemistry Letters*, **4**, 2821–4.

Sofia, M. (1996) Generation of Oligosaccharide and glycoconjugate libraries for drug discovery, *Drug Discovery Today*, **1**, 27–36.

Storer, R.C. (1996) Solution phase synthesis in combinatorial chemistry: applications in drug discovery, *Drug Discovery Today*, **1**, 248–54.

Strocker, A.M., Keating, T.A, Tempest, P.A. and Armstrong, R.W. (1996) Use of a convertible isocyanide for generation of Ugi reaction derivatives on solid support: synthesis of α-acylaminoesters and pyrroles, *Tetrahedron Letters*, **37**, 1149–52.

Tam, J.P. and Lu, Y. (1995) Coupling difficulty associated with interchain clustering and phase transition in solid phase peptide synthesis, *Journal of the American Chemical Society*, **117**, 12058–63.

Terrett, N.K., Gardner, M., Gordon, D.W., Kobylecki, R.J. and Steele, J. (1995) Combinatorial synthesis – the design of compound libraries and their application to drug discovery, *Tetrahedron*, **51**, 8135–73.

Thompson, L.A. and Ellman, J.E. (1996), Synthesis and applications of small molecule libraries, *Chemical Reviews*, **96**, 555–600.

Tomczuk, B.E., Gowravaram, M.R., Johnson, J.S., Cook, E.R., Ghose, A.K., Mathiowetz, A.M., Spurlino, J.C., Rubin, B., Smith, D.L., Pulvino, T.A. and Wahl, R.C. (1995) Hydroxamate inhibitors of the matrix metalloproteinases (MMPs) containing novel P_1' heteroatom based modifications, *Bioorganic and Medicinal Chemistry Letters*, **5**, 343–6.

Turner, N.J. (1996) presented at *Combinatorial Chemistry Symposium 2*, University of Exeter, UK, July.

Uozumi, Y., Danjo, H. and Hayashi, T. (1997) New amphiphilic palladium–phosphine complexes bound to solid supports: preparation and use for catalytic allylic substitution in aqueous media, *Tetrahedron Letters*, **38**, 3557–60.

Valero, M., Giralt, E. and Andreu, D. (1996) Solid phase-mediated cyclization of head–to–tail peptides: problems associated with side chain clustering, *Tetrahedron Letters*, **37**.

Virgilio, A.A. and Ellman, J.A. (1994) Simultaneous solid phase synthesis of β-turn mimetics incorporating side-chain functionality, *Journal of the American Chemical Society*, **116**, 11580–1.

Wahl, R.C., Pulvino, T.A., Mathiowetz, A.M., Ghose, A.K., Johnson, J.S., Delecki, D., Cook, E.R., Gainor, J.A., Gowravaram, M.R. and Tomczuk, B.E. (1995) Hydroxamate inhibitors of human gelatinase B (92 kDa), *Bioorganic and Medicinal Chemistry Letters*, **5**, 349–52.

Wentworth, P., Vandersteen, A.M. and Janda, K.D. (1997) Poly(ethylene glycol) (PEG) as a reagent support: the preparation and utility of a PEG–triarylphosphine conjugate in liquid phase organic synthesis (LPOS), *Journal of the Chemical Society: Chemical Communications*, 759–60.

White, M.A. (1992) in *Solid Supports and Catalysis in Organic Synthesis* (ed. K. Smith), Ellis Horwood, Chichester, p. 225, and references therein.

Wilson, S.R. and Czarnik, A.W. (1998) *Combinatorial Chemistry: Synthesis and Application*, Wiley, Chichester.

Wilson-Lingardo, L., Davis, P.W., Ecker, D.J., Hérbert, N., Acevedo, O., Sprankle, K., Brennan, T., Schwarcz, L., Freier, S.M. and Wyatt, J.R. (1996) Deconvolution of combinatorial libraries for drug discovery: experimental comparison of pooling strategies, *Journal of Medicinal Chemistry*, **39**, 2720–6.

Wright, P., Lloyd, D., Rapp, W. and Andrus, A. (1993) Large scale synthesis of oligonucleotide *via* phosphoramidite nucleosides and high-loaded polystyrene support, *Tetrahedron Letters*, **34**, 3373–6.

Yan, B., Fell, J.B. and Kumaravel, B. (1996) Progression of organic reactions on resin supports monitored by single bead FTIR microspectroscopy, *Journal of Organic Chemistry*, **61**, 7467–72.

Yan, L., Taylor, C.M., Goodnow, R. and Kahne, D. (1994) Glycosylation on the Merrifield resin using anomeric sulfoxides, *Journal of the American Chemical Society*, **116**, 6953–4.

Ye, Q.-Z., Johnson, L.L., Nordan, I., Hupe, D. and Hupe, L. (1994) A recombinant human stromelysin catalytic domain identifying tryptophan derivatives as stromelysin inhibitors, *Journal of Medicinal Chemistry*, **37**, 206–9.

Yeh, L.A., Chen, J., Baculi, F., Gingrich, D.E., and Shen, T.Y. (1995) Inhibition of metalloproteinase by futoeneone derivatives, *Bioorganic and Medicinal Chemistry Letters*, **5**, 1637–42.

Zhang, H.-C. and Maryanoff, B.E. (1997) Construction of indole and benzofuran systems on the solid phase *via* palladium-mediated cyclization, *Journal of Organic Chemistry*, **62**, 1804–9.

10

New Methods for Structure-based *De Novo* Drug Design

Bohdan Waszkowycz

10.1 THE ROLE OF *DE NOVO* DESIGN IN DRUG DISCOVERY

10.1.1 Introduction

Drug discovery traditionally has relied more on serendipity than on rational design, especially in the early stages of lead molecule identification. The random screening of natural products or pharmaceutical companies' chemical libraries has accelerated markedly in recent years with the development of high throughput assay techniques. Combinatorial chemistry has sought to supply the need for ever-expanding compound libraries by offering the potential of accessing very large libraries of peptidic and (increasingly now) non-peptidic organic compounds (Gallop *et al.*, 1994; Gordon *et al.*, 1994; see also Chapter 9 by Shuttleworth). At present, the optimization of a lead compound into a pre-clinical and ultimately a clinical drug candidate follows a more traditional medicinal chemistry approach of exploring structure–activity relationships by methodical synthesis of analogues, guided by the expertise of the medicinal chemist.

However, there has always been interest in pursuing a rational approach to drug discovery, driven by the opportunity to accelerate the optimization stage or to access greater chemical novelty. In the lead optimization phase, the study of quantitative structure–activity relationships (QSAR), led by the disciplines of molecular modelling and physical property estimation, has had proven value in rationalizing the structural and physico-chemical properties leading to activity. Similarly, the concept of the pharmacophore has become commonplace as a representation of the key features which characterize affinity and/or

Advances in Drug Discovery Techniques. Edited by Alan L. Harvey © 1998 John Wiley & Sons Ltd.
ISBN 0 471 97509 5

selectivity in a ligand, and which are recognized by a specific biological receptor molecule possessing complementary properties.

The importance of QSAR and pharmacophore elucidation reflects the fact that most drug discovery programmes are directed towards biological targets which are poorly characterized. While a broad range of receptors and enzymes as therapeutic targets is being pursued, only for a minority of targets is there sufficient structural information available to allow direct or structure-based drug design (SBDD) (Bohacek, McMartin and Guida, 1996). The goal of SBDD is to design ligands which are complementary in properties to the binding site, e.g., in terms of shape, hydrogen bonding and lipophilicity. Maximizing the complementarity of the ligand will lead to improved binding affinity and receptor specificity, which are important properties (though by no means the only ones) to optimize in the pursuit of enhanced efficacy and safety.

In order to design a ligand which matches the binding site, it is essential that the 3D structure of the target receptor is known in atomic detail. In practice, this requires crystallization of the target protein and determination of the 3D structure by x-ray diffraction. (For small proteins, an alternative route is NMR spectroscopy of the protein in solution.) In ideal cases it may be possible to co-crystallize a known ligand or inhibitor into the binding site, and this approach is being pursued increasingly in drug discovery programmes. It gives confirmation of the binding mode of the ligand and the nature of the interactions made with the receptor, and reveals the extent to which the receptor structure deforms to accommodate the ligand. This becomes a very powerful tool for further optimization of the ligand, e.g., to identify where contacts are sub-optimal or where extra contacts can be made, or to suggest substitution positions where the physical properties of the ligand may be modified without interfering with receptor binding. While such inferences and modifications can be made manually, there is also value in using molecular modelling software to assist this process, e.g., to perform a more systematic and quantitative analysis. This idea of using (partially) automated molecular design tools has been developed over a number of years now, and is commonly referred to as *de novo* design (Böhm, 1993; Lewis and Leach, 1994).

10.1.2 The importance of structural chemistry

To date, the main reason for the relatively limited application of SBDD has been the lack of structural data across the full range of therapeutic targets. There are many technical reasons why successful x-ray crystallography of a desired protein often proves difficult, if not in some cases impossible. The protein must be extracted from natural sources or cloned and expressed in suitable quantities, and must yield crystals of satisfactory quality following what can be a fairly hit-and-miss crystallization process. Many families of proteins prove difficult to crystallize because they are naturally membrane-bound or multimeric, or otherwise unstable under the crystallization conditions. However, these technical aspects are continually being addressed: recent advances in bioinformatics and molecular biology, for example, have vastly accelerated the characterization and expression of proteins. As a result, recent years have witnessed an exponential rise in the number of protein crystal structures available in the public domain. For instance, the current release of the Brookhaven Protein Data Bank numbers over 5000 structures (mostly protein), with new structures arriving at the rate of over 1000 per year, compared with less than 100 per year a decade ago. However, the majority of crystal structures are for soluble enzymes, while many classes of therapeutic targets of interest to the medicinal chemist (such as membrane-bound receptors) remain intractable.

It remains true that obtaining a crystal structure of a new protein may take anything from months to years of work, which clearly is a barrier to expecting SBDD to be applicable to any desired target. However, it is true also that, having once achieved crystallization of a target protein, it is (in favourable cases) possible to solve a large number of complexes with different ligands in a very short timescale. Recent work with HIV protease demonstrates clearly the value of this approach in determining the binding modes and SAR of structurally diverse classes of inhibitors, and the potential for exploiting such information in rational design (Bohacek, McMartin and Guida, 1996). It is estimated that by now many hundreds of HIV protease–inhibitor complexes have been determined within pharmaceutical companies, with only a fraction of that number in the public domain. Some representative successful examples of SBDD are given in Table 10.1.

The steady increase in the number of resolved receptor–ligand complexes is important not only for the direct impact in drug discovery directed at a given receptor, but also for the wealth of data accumulated in the area of molecular recognition. This has facilitated the analysis of the key features of receptor–ligand recognition, leading to an increasingly firm rationale for the underlying mechanisms. To a great extent, molecular recognition is now understandable in terms of fundamental physical processes, including, e.g., the concepts of complementarity of molecular properties (shape, electrostatic potential, hydrophobicity/polarity), the importance of induced fit and strain, and the enthalpic and entropic contributions of intermolecular interactions and solvent reorganization (Böhm and Klebe, 1996). This large database of information serves also to improve the quality of software across the whole discipline of computer-aided molecular design (CAMD), providing more reliable and quantitative algorithms and models in such fields as molecular simulation, docking, definition of molecular similarity and diversity, and prediction of binding affinity.

Table 10.1 Examples of enzyme inhibitor discovery involving some contribution of structure-based drug design. For references see Bohacek, McMartin and Guida (1996) and Böhm and Klebe (1996)

Target enzyme	*Clinical application*	*Status*	*Company*
Carbonic anhydrase	Glaucoma	Launched	Merck
HIV protease	AIDS	Phase II/launched	Roche, Merck, Abbott, Vertex
Purine nucleoside phosphorylase (PNP)	T-cell lymphoma, psoriasis	Phase II/III	Biocryst
Thymidylate synthase	Cancer	Phase I/II	Agouron
Thrombin	Thrombosis	Phase II	Roche
Influenza neuraminidase	Influenza	Phase II	Glaxo-Wellcome
Cyclophilin-calcineurin	Immunosuppression	Preclinical	Vertex
Interleukin-1beta converting enzyme (ICE)	Inflammation	Preclinical	Vertex, Chiroscience, Hoechst-Marion-Roussel, Novartis

10.1.3 *De novo* design

Given the success of structural chemistry data in providing a rational focus to lead optimization, it is reasonable to explore the value of SBDD in the more demanding task of lead discovery, i.e., identifying a ligand of novel chemistry. This must be inherently a far more risky process than simple derivatization of a known lead. In analogue synthesis, the chemist can reasonably expect to achieve a certain level of activity, and may quickly rationalize why activity has been increased or abolished. In contrast, in *de novo* design the problem is that there are many properties of the ligand which have to be predicted correctly if the ligand is to prove a hit in the primary screen: e.g., are the desired binding interactions accessible in reality?; is the predicted ligand conformation achievable?; are the physico-chemical properties of the ligand suitable?; and is the designed molecule synthesizable? The more novelty that is introduced in any of these areas, then the greater the likelihood of a prediction going badly awry.

However, there are many reasons why *de novo* design can make a useful contribution to drug discovery. Much of the drug discovery process is driven by the need for some degree of chemical novelty. For example, there may be a limit on how far a given lead can be optimized for a receptor, and hence a significant increase in affinity or selectivity arises only by moving to a new lead series. Many drug candidates fail in the development stage (e.g., on issues such as pharmacokinetics and toxicity) where again a significant improvement in properties may not be achievable by minor modification. There are also the emerging new targets, for which no satisfactory lead molecule exists, and all design is therefore effectively *de novo*. Finally, there is the important issue of patentability, of needing some degree of novelty in order to bypass a protected chemical series. Lead discovery admittedly continues to be pursued primarily by random screening; see Chapter 3 by Hill. However, there is a clear opportunity also for a rational approach, especially in those cases where the receptor structure has been well characterized and where it is obvious that this knowledge should be exploited to some degree to influence the choice of molecule to be screened or synthesized.

Clearly, *de novo* design can be performed manually by the experienced medicinal chemist, with no computational tools more sophisticated than computer graphics to display the structure of the receptor. Is there then a need for *de novo* design software to perform the design and evaluation of ligands automatically? The potential offered by *de novo* design software is to be able to generate novel ideas objectively, without the limitations or bias imposed by an individual's personal insight and expertise, and to explore and evaluate a large number of possible solutions. (Ideally, the software would retain the possibility to bias or direct the solutions where this is deemed appropriate.) Novelty can be achieved because a computational algorithm can consider all viable solutions systematically. For example, in a typical fragment-based approach to molecule building, the number of combinations of fragments (and their 3D orientation and conformation) increases exponentially, so that the human designer can build and evaluate only a fraction of the solutions which are feasible. A typical *de novo* design program will build and analyse perhaps tens of thousands of solutions or partial solutions in a single run.

10.1.4 Benefits and limitations of *de novo* design

With regards to *de novo* design, the main concern of the medicinal chemist is whether the solutions are sensible enough to be worth the effort of synthesis. The quality of the design is judged in terms of chemical properties (e.g., physico-chemical properties, synthetic feasibility) and modelled properties (e.g., binding mode, binding affinity). Accurate prediction of each of these represents a considerable challenge to computational chemistry: there are no computational methods which are 100% accurate for any of these problems. To many medicinal chemists this lack of absolute accuracy may be enough to undermine their confidence in the worth of *de novo* design. However, this is to miss the point: *de novo* design is a technology which can provide a unique and valuable insight into the difficult task of rational ligand design, and which has a role to play alongside many other approaches within a multidisciplinary drug discovery programme.

De novo design is a tool intended to assist with the analysis of a specific problem: to understand the characteristics of a given receptor, to predict in general terms what kinds of chemistry should be looked for in a ligand, and then more specifically to identify chemically reasonable (but nevertheless novel) designs which are realistic examples of the type of ligand which should be pursued. The real benefit of *de novo* design is as an ideas generator, which can be as focused or as broad as the problem demands. However, it is naïve to think of *de novo* design software as a black box which produces highly active lead molecules with no further effort on the part of the modeller or medicinal chemist. Each library of designs inevitably will require development to some degree. For example, the modeller will decide to add further constraints to the design run in order to explore more fully a solution which looked particularly attractive, or to prevent the generation of unfavoured solutions. The medicinal chemist will want to translate or modify the chemistry into one which is synthetically more accessible but which retains the same binding interactions. Hence *de novo* design is a tool which needs to be used intelligently and creatively, and needs to be driven by the insight and imagination of the modeller and medicinal chemist.

It is worth making some comparisons with the complementary technology of searching databases of 2D or 3D chemical structures for compounds which are deemed similar to a known lead molecule or which match a pharmacophore generated from a known receptor. This has the advantage over *de novo* design that all hits are real molecules rather than hypothetical molecules, and may be accessible instantly from an in-house compound library. The main disadvantage is that, even where the library numbers hundreds of thousands of compounds, only very limited chemical diversity may be present, and the chances of finding a set of good ligands may be relatively poor. Another potential problem with database searching is that searching against a simple (e.g., 3 or 4 point) pharmacophore typically will yield a vast number of hits. Simple molecular modelling will demonstrate that the majority of ligands which fit a loosely defined pharmacophore will not actually fit well the much more constrained environment of a receptor binding site. Hence there is value in further processing a hit list from a database search, using a variety of CAMD techniques.

Therefore *de novo* design is particularly useful in situations where exploration of diversity and novelty is considered important, whether this is part of a lead identification or lead optimization programme. *De novo* design can cover a broad spectrum of applications, from the limited modification of known ligands to the design of completely novel chemistries, and indeed progress continues to be made in situations where no receptor structure is available. In the latter case, *de novo* design can be driven to mimic a set of active ligands, with further constraints given by QSAR data or pseudoreceptor models.

Some successful examples of the use of *de novo* design software are given in Table 10.2.

Table 10.2 Examples of published *de novo* design applications

Application	*Program*	*Reference*
Design of peptidic renin inhibitors	GROW	Moon and Howe (1991)
Novel inhibitors of FKBP-12	LUDI	Babine *et al.* (1995)
Predition of binding mode of peptidic HIV protease inhibitor	MCSS	Caflisch, Miranker and Karplus (1993)
Optimization of non-peptidic HIV protease inhibitor	MCDNLG	Gehlhaar *et al.* (1995)
Optimization of known HSF-PLA2 inhibitors	LUDI	Pisabarro *et al.* (1994)
Library design for thrombin inhibition	PRO_SELECT	Murray *et al.* (1997) Young *et al.* (1998)

10.2 GENERAL METHODS AND CONSIDERATIONS IN *DE NOVO* DESIGN

10.2.1 Overview of *de novo* design methods

De novo design is still very much an evolving discipline within CAMD, and there is as yet no algorithm or procedure acknowledged as being the 'best' way to generate novel molecular designs of good diversity and chemical sense. Available software packages pursue a range of different approaches, each with its own advantages and drawbacks. Most packages attempt to mimic and automate the route that a modeller or medicinal chemist would take in pursuing manual design to a receptor structure. The overall strategy can be broken down into the following key stages, which may be addressed in a single software package, or, more usually, may require access to several different packages, together with some degree of interpretation and manipulation by an expert chemist or modeller:

- generation of receptor models,
- characterization of the binding site,
- definition of specific design objectives or constraints to focus the structure building run,
- selection of a database of molecular fragments to be used for structure building,
- assembly of molecular fragments in the binding site,
- first-line scoring and ranking of all solutions,
- more detailed evaluation of favoured designs and prioritization for re-design or synthesis.

De novo design software has originated mostly from modelling groups within academia or the pharmaceutical industry. Several packages are now available commercially from the major CAMD software suppliers. A brief (and not exhaustive) list of programs described in the literature is given in Table 10.3.

10.2.2 Necessary data

The starting-point for *de novo* SBDD is a 3D structure of the receptor of suitable quality (to a resolution of 2 Å or less), and ideally with a ligand complexed in the binding site. There are many problems which may limit the usefulness of the x-ray data, e.g., there may be poorly resolved surface loops, or ambiguous density from a weakly bound ligand. Even for a well resolved structure, it should be remembered that the structure observed is only one of perhaps a number of equally valid conformations, and that some structural

Table 10.3 Examples of *de novo* design and related programs

Program	*Key features*	*Reference*
GROW	Sequential growth of oligopeptides from amino acid fragments guided by binding energy estimation	Moon and Howe (1991)
LEGEND	Atom-by-atom approach to general organic design	Nishibata and Itai (1991)
LUDI	General organic design directed by interaction sites: initial targeting of key pharmacophoric sites via a large fragment library followed by bridging	Böhm (1992)
MCSS/HOOK	Linking of fragments initially docked via molecular dynamics simulation	Caflisch, Miranker and Karplus (1993)
NEWLEAD	Construction of linkers/scaffolds to bridge pre-placed pharmacophoric fragments	Tschinke and Cohen (1993)
SPROUT	Carbon skeleton construction, followed by heteroatom replacement. Interaction sites mapped out using HIPPO	Gillet *et al.* (1993)
GenStar/ GROUPBUILD	Energy-directed sequential growth of general organics, using an atom or fragment approach, respectively	Rotstein and Murcko (1993a,b)
PRO_LIGAND	Interaction site-directed structure generation for general organics and peptides. Allows for a variety of methods: direct and indirect drug design, sequential growth and fragment bridging, conformationally rigid or flexible fragment placing, etc.	Waszkowycz *et al.* (1994) Clark *et al.* (1995) Frenkel *et al.* (1995)

details may be an artefact of crystal packing. The modeller must decide how to treat conformationally flexible sidechains, and whether to include or exclude water molecules buried in the binding site. The absence of a ligand may perturb the structure of the binding site, as may the presence of a sub-optimal ligand.

If a crystal structure is unavailable, an alternative would be to derive a model based on structurally related proteins. The problem with this approach is that models rarely achieve the accuracy necessary for acceptable *de novo* design: a few ångstroms deviation in the position of an amino acid residue often will substantially affect the size and nature of a binding pocket. A model obtained from a highly homologous protein may be acceptable (where the binding site differs only by conservative substitutions), but if variable surface loops or insertions and deletions form part of the binding site, then such models must be treated with caution.

The availability of a co-crystallized receptor–ligand complex is not essential but can prove invaluable in delineating the extent of the binding site and the key binding interactions, and confirming that the observed conformation of the active site does indeed allow ligand binding. Obviously, it is a very secure starting point if one's objective is to generate modifications of the lead compound rather than novel designs. SAR data on a number of active and inactive analogues also are valuable, even if crystal data on those analogues are not available. In such a case one may deduce (following some modelling) a consensus of features which appear essential or detrimental to binding, as not all the contacts observed between ligand and receptor may be essential to binding, or some contacts may have a greater influence than others on the observed affinity. Such data can be important in focusing the *de novo* design process: rather than have a ligand bind anywhere in the binding site, one can add a constraint that, say, an interaction with a specific amino acid must be achieved. It may be possible also to deduce information on the general physical properties desired for the particular class of inhibitor, e.g., a certain range of molecular weight or log *P*. In the ideal case, the training set would allow the construction of a robust QSAR model, with the advantage that a more accurate prediction of binding affinity may be made for new designs.

Where the receptor structure is unknown, progress generally is made by a classical medicinal chemistry route, generating analogues in order to define the SAR and eventually to rationalize a pharmacophore leading to activity. Relatively little has been reported for attempting *de novo* design similar to a known ligand or similar to a pharmacophore (as opposed to being complementary to a receptor). The problem in designing in the absence of a receptor structure is that a pharmacophore or even a modelled pseudoreceptor is a far less well defined object than a true receptor. Hence the process of ligand design will be less constrained, leading to more diverse solutions being generated which cannot be evaluated and scored with the same accuracy or confidence as in the case of SBDD.

10.2.3 Characterization of the binding site

Several *de novo* design programs build directly into the binding site, with fragments selected randomly from a database, and then scored in terms of an empirical (molecular mechanics) binding energy. For these programs, no preliminary analysis or characterization of the binding site is required, other than to define the boundaries of the site, and, if desired, to define a seed fragment as a starting point for structure generation.

By contrast, other programs calculate and map out various properties of the binding site which then are used to guide the selection and positioning of molecular fragments. An example of how this is implemented is in the use of 'interaction sites' in the program LUDI (Böhm, 1992) and our program PRO_LIGAND (Clark *et al*., 1995). Interaction sites represent regions within the binding site where it is predicted that the receptor may offer good hydrogen bonding or lipophilic contacts with a putative ligand. These can be displayed as volumes or as discrete vectors or points. Interaction sites usually are generated geometrically around individual amino acid residues, using a database of observed ligand–receptor contact geometries. For example, a single polar receptor atom may give rise to tens of hydrogen bonding sites, representing the range of angles and distances at which an acceptable idealized hydrogen bond may be formed. In effect the collection of sites represents a 3D map of the binding site, describing the location of contacts accessible to a ligand. It can be thought of as a pharmacophore, albeit a very detailed one (i.e., with typically hundreds of sites defined rather than the handful of sites in a classical pharmacophore). For convenience, a collection of interaction sites is referred to as a 'design model' in PRO_LIGAND.

The design model serves as a template to identify fragments within a database which possess matching functionality in a similar 3D orientation. Suitable fragments are then docked directly on to the appropriate subset of interaction sites using standard superimposition methods. The fragment database comprises either individual chemical moieties or complete molecules, with hydrogen bonding and lipophilic functionality identified and labelled. In our program PRO_LIGAND, a graph theory approach (i.e, akin to methods used routinely in chemical database searching) is used to decide whether a given fragment in the database is capable of fitting on to any subset of sites in the design model. Clearly, any given fragment may be capable of binding to several subsites and in several orientations, but all these alternative placements can be identified by the graph theory algorithm.

The design model has the advantage of giving the modeller more control in directing the building process. For example, growth may be localized to a narrow region of the receptor simply by pruning the number of interaction sites using appropriate graphical tools.

There are other, generally more computationally costly, approaches to explore and characterize the binding site. The program GRID (Goodford,

1985) calculates a 3D potential energy field for a probe atom or moiety interacting with the receptor, and thereby identifies regions where a good binding energy is expected. The program MCSS (Caflisch, Miranker and Karplus, 1993) performs molecular dynamics simulations on small fragments complexed to the receptor, in order to identify the regions and orientations at which a given chemical fragment will optimally bind. The information is then used by the program HOOK, which attempts to link together the most favoured fragments and orientations to form more realistic ligands.

10.2.4 Structure generation

The process of structure generation is the stage at which the most significant differences arise among individual programs. At its simplest, the problem is one of how to build novel chemical structures efficiently within the constraint of the receptor envelope or the design model. Some of the more important considerations are now described briefly.

(a) Building-block definition

The basic building-block for constructing designs may be a complete molecule, a chemically simpler small fragment, or an individual atom. Accessing a database of known compounds ensures that chemically sensible designs are achieved, but at the expense of structural diversity. Clearly, the other extreme, of constructing molecules on an atom-by-atom basis, guarantees unlimited chemical diversity, but there will be problems in ensuring synthetic accessibility. Therefore, many workers have used a fragment-based approach as a pragmatic compromise, where fragments may typically be small functional units, like small alkanes, rings, amines, alcohols, etc. Even with relatively small fragment libraries, very good chemical diversity is achieved. Synthetic feasibility can be controlled partially by defining rules to limit the ways in which fragments can be joined together. It is important to ensure that it is straightforward for the modeller to add fragments to the database (e.g., if a particular scaffold chemistry is desired), and also to prioritize the way in which classes of fragments are selected; this can be useful in encouraging the construction of sensible chemistries.

(b) Approaches to molecule building

There are two common extremes. A molecule can be 'grown' progressively by adding one fragment at a time to a starting fragment, thereby ensuring that the solution is a complete molecule at all times. Alternatively, one can place fragments in an 'outside-in' mode, such that the most favoured binding sites are filled first. This has the advantage that there is perhaps more chance that such sites are optimally filled, but it then relies on finding suitable bridging or linking chemistries to connect the fragments together into a complete molecule, and this can be a difficult process. However, these approaches should not be thought of as mutually exclusive, and in PRO_LIGAND we have implemented an open-ended approach whereby a combination or intermediate strategy can be defined easily.

(c) Conformational flexibility

When trying to place medium-to-large fragments, some method is needed to take account of conformational flexibility, as otherwise many potential dockings will be missed. The problem partly disappears if a library of small fragments is used, such that no rotatable bonds are present within a fragment, but this unduly restricts the nature of the fragment library. One way to address the general problem is to perform torsional rotation during the docking stage. This can become computationally expensive: some method of conformational searching is required in order to fit the fragment on to the design model, possibly replicated several times in order to avoid trapping in local minima, while monitoring internal strain energy. An alternative approach is to predefine a fragment database in which each fragment is represented by several conformers, which then are fitted rigidly on to the design model. This route has the advantage that, by pre-defining the conformer library, one may ensure that a good selection of realistic conformations is present. The run-time execution of the program is not delayed other than for the problem of having to search what is in effect a larger fragment database. For example, several hundred conformers may be required for an amino acid fragment.

(d) Searching algorithms

A fragment-based building procedure represents a very large combinatorial problem: many fragments need to be evaluated at each stage, in many orientations. Therefore a pragmatic searching strategy usually is required to identify good solutions without having to search every node of the decision tree. To attempt to identify the best fragment at each stage of the building process is costly in time and may not be the best strategy in the long run: it may be better to accept some sub-optimal partial solutions in order to get to a better final solution. Therefore we prefer a strategy of accepting a fragment as soon as a plausible solution is found (within defined limits) rather than to continue to search for better solutions. This has the benefit that one can quickly generate a large number of solutions (typically hundreds or thousands), with high diversity but variable quality. Then the problem becomes one of scoring the solutions so that the user needs to inspect only the best quality solutions.

(e) Structure optimization

This can be part of the process of structure building or can be an additional post-processing stage. The quality of the designs can be improved in several ways: by adding extra functionality, such as heteroatom replacements, rigidifying flexible structures by ring formation, or screening for unwanted properties. As one example, a 2D search for moieties considered unsuitable on the grounds of synthetic feasibility or toxicity is straightforward to implement if one has defined a database of unwanted substructures. *De novo* design should be thought of as an iterative procedure, so one useful route to optimization is to select a favoured design and to use that as a starting point for additional growth or substitution in subsequent structure building runs. A more interesting approach is to attempt to optimize more objectively a set of solutions, for example using a genetic algorithm. We have found this to be useful when used in conjunction with PRO_LIGAND, which tends to produce large numbers of solutions of variable quality. These can serve as an initial 'population' for the optimization algorithm: in our implementation, solutions are randomly 'bred' (i.e., molecules are split and rejoined across bonds in different combinations) so that the population is progressively optimized as favoured substructures are identified (Westhead *et al.*, 1995).

10.2.5 Scoring and evaluation

It is essential that a method is available for ranking the quality of the designed structures in terms of properties which are meaningful to both the modeller and medicinal chemist. The accuracy of the scoring procedure is of particular importance when it is remembered that a *de novo* design program may be generating thousands of solutions which the modeller will not be able to inspect individually, and hence if they are to be of any use, they must be scored and ranked realistically. Generally the score is a prediction of binding affinity: for example, a simple estimate of receptor–ligand enthalpy of binding commonly is calculated within a molecular mechanics formalism, with perhaps an estimate of ligand strain. The problem here is that a fast approach is needed, as typically many thousands of binding affinity evaluations need to be performed. However, it is well known that the accuracy of a simple single-point enthalpy calculation generally is poor, and does not correlate well with the free energy of binding. The most accurate force-field methods of estimating free energy, based on extensive molecular dynamics or Monte Carlo simulation, are very time consuming and impossible to implement within a *de novo* design strategy where many ligands need to be assessed relatively rapidly.

As an alternative to molecular mechanics enthalpies, a scoring function can be defined based simply on the type and nature of the interaction sites hit by the ligand, complemented by penalties for introducing rotatable bonds, chiral centres or other structural features considered undesirable. The advantage of such a scoring function is that it can be modified or weighted by the user to emphasize a particular type of solution (e.g., good lipophiles rather than hydrogen bond donors may be desired). The disadvantage is that the score has perhaps little correlation with actual binding affinity. Hence it becomes more important that a more robust prediction of affinity is applied to the final solutions. If there are large numbers of solutions to be scored, speed remains an issue. The most promising approach at present appears to be the use of empirically derived scoring functions like that due to Böhm (1994). Features such as hydrogen bonding and lipophilic contacts between ligand and receptor are measured and scored via a function that has been calibrated to reproduce the observed binding affinity of a large database of diverse crystallographically determined receptor–ligand complexes. Compared with direct simulation methods, this has the benefit of speed and transferability across diverse receptor–ligand chemistries, although necessarily accuracy is limited (to at best 1–2 orders of magnitude in binding affinity). A major source of error is the assumption that the *de novo* design program has generated a design in an accessible low energy conformation and has docked it in the most favoured subsite. Therefore at best all scoring functions serve only to highlight the most promising solutions and must not be applied uncritically.

Clearly, prediction of binding affinity is only the first screen to be applied in ranking solutions from the *de novo* design program. Before selecting candidates for re-design or synthesis, other properties should be examined, including:

- synthetic feasibility: this can be assessed computationally (e.g., using programs such as CAESA (Gillet *et al.*, 1995)) but is more likely to rely on human expertise;
- general physico-chemical properties, such as solubility, hydrophobicity (log P), acidity/basicity (pK_a): to a first approximation, many of these can be calculated in a range of software packages, and can be especially useful when it is known that a particular range of values is relevant for a given therapeutic target;
- general biological properties, such as metabolism, pharmacokinetic profile, toxicity: these are always of critical importance for a drug but are difficult problems to address computationally, although progress continues to be made, for example in compiling appropriate databases.

10.3 SYNTHETIC ACCESSIBILITY WITHIN PRO_SELECT

10.3.1 Introduction

The goal of *de novo* design is the generation of many diverse ligands for a given target. However, given that it is not possible to predict binding mode and binding affinity completely reliably, lead discovery programmes cannot run the risk of committing excessive resource to the synthesis of novel molecules with perhaps a small chance of success. Therefore high priority must be given to searching databases of available reagents or in-house libraries to identify suitable mimics of a designed ligand.

In order to increase the likelihood of success, it is important that novel designs are accessible by reasonably facile synthetic routes, and that a small library of analogues is synthesized, rather than just the parent compound. There are several ways of constraining the structure generation process to improve synthetic accessibility, e.g., specific substitution positions on a fragment can be favoured or excluded. Alternatively, the design problem may be restricted to the elaboration of a desired scaffold, so that a set of solutions will be amenable by a single synthetic route. While these techniques can restrict the degree of chemical diversity, this is a necessary compromise if the modeller desires to see a large number of designs synthesized and tested.

We recently described a process for lead discovery and optimization, PRO_SELECT, which is a further development of the above ideas (Murray *et al.*, 1997; Young *et al.*, 1998). The objective was to focus the *de novo* design program such that a library of designs was produced which was compatible with a single defined reaction scheme. It is important to note that a library here refers to perhaps tens or hundreds of synthetic targets rather than the handful of targets which is a typical outcome of a *de novo* design exercise. The synthesis of a moderately sized library is important: it recognizes the fact that all predictions are limited in accuracy, and hence success is more likely when a reasonable number of designs are synthesized and a range of molecular properties represented. In addition, the design of a library of broadly homologous structures may allow some valid SAR deductions to be drawn at a relatively early stage of a drug discovery project. Again this is rarely possible if only a small number of chemically diverse *de novo* designs are synthesized.

The synthesis of a large library of homologous compounds is achieved efficiently by following a combinatorial chemistry approach, i.e., by identifying a synthesis route which can yield an array of compounds which share a common template or scaffold chemistry, and with a diverse series of substituents at one or more substitution positions. In order to be practically useful, the synthesis route must be sufficiently robust and there must be reagents of suitable diversity readily available. Normally, combinatorial chemistry applications are directed towards very large libraries (for non-peptidic organics, libraries of thousands of compounds are now being synthesized). Within PRO_SELECT we preferred to make smaller, more targeted libraries (perhaps only tens of compounds in a first-pass library, expanding to hundreds in further iterations), and to make single, relatively pure compounds via parallel synthesis routes, rather than make crude mixtures of compounds, as typical of some combinatorial chemistry applications. This approach has the advantage that the modeller has access to good quality data on well defined compounds, and that the synthetic chemistry is not constrained to a single high throughput route, i.e., some modification of the route can be tolerated in order to accommodate more diverse functionality or properties.

Within PRO_SELECT, synthetic feasibility is ensured by constraining the structure generation protocol to design a virtual library based on a defined synthetic route and using only readily available reagents. As combinatorial chemistry is capable of producing libraries of perhaps millions of compounds, a key problem in library design is to select a subset of designs to be synthesized. Although there are several approaches in the literature which consider library design, generally by means of similarity or diversity analysis of the reagent lists, we preferred to select favoured substituents by means of screening against the receptor model, and thereby identify substituents which are predicted to be complementary to the properties of the receptor binding site. There is an important difference between this approach and more general diversity analysis of reagent lists. Within PRO_SELECT it is recognized that a very narrow range of properties is required to allow binding to the receptor, but within that range of properties we want to identify as broad a diversity of molecules as possible. Thus the library will not be as 'diverse' as a general combinatorial chemistry library but rather is focused or targeted towards the known properties of the receptor.

It is not possible to screen computationally a large combinatorial library against a receptor model in order to assess binding modes and binding affinity objectively. Even though some docking methods can now cope with tens of thousands of compounds, very fast and hence very approximate methods must be used, resulting in many false positives and negatives. Therefore an important idea in PRO_SELECT is that the computational screening is made tractable by successively breaking down the problem. First, each substitution position is analysed in isolation from the rest. This enables large substituent lists at each position (say thousands of reagents) to be handled in an acceptable time. A subset of the most favoured substituents at each position is selected for full combinatorial enumeration and further computational analysis. In this way the enumerated virtual library never becomes unmanageable in size.

10.3.2 Method

An overview of the methodology is given in Figure 10.1. The first stage, as for *de novo* design, is the generation of suitable molecular models of the receptor binding site. Analysis and characterization of the binding site (by generation of design models) is performed in order to assess the general properties required in the library of designs. At this point molecular modelling will play an important role in evaluating the suitability of one or more suggested template chemistries: in deciding where a template is likely to bind and how it is oriented in the receptor. Note that templates usually are not modelled in isolation but rather with plausible substituents attached. Clearly, this process is more secure if there is experimental evidence for the binding site of the template. The

Characterize receptor binding site

Identify appropriate template chemistry and reaction scheme

Derive models of receptor–template complex

Generate pharmacophores and design models

Generate substituent lists from reagent database

Perform 3D receptor-based substituent screen

Eliminate unfavoured substituents on basis of predicted properties

Enumerate virtual library of favoured substituents

Perform receptor-based screen of virtual library

Select candidates for synthesis

Figure 10.1 Overview of the virtual library generation and screening paradigm used within PRO_SELECT (Murray *et al.*, 1997). The tasks in bold are performed using proprietary software, while other tasks employ a variety of commercial modelling and database software.

accessible substitution positions on the template must be directed towards areas of the receptor which are considered to be worthwhile subsites for binding.

An example is the generation of a library of thrombin inhibitors based on a central proline template. PPACK (D-Phe-Pro-Arg-chloromethylketone) is a well known peptidic inhibitor of thrombin. It was desired to generate a library of compounds of the type A-Pro-B in which the A position broadly explores mimics of the D-Phe amino acid and the B position explores mimics of Arg. In this case, positioning of the template in the active site of thrombin was secure, being guided by the available x-ray data of the thrombin–PPACK complex, and it was known that substituents at the N terminus and C terminus of proline would be directed primarily towards the large aryl hydrophobic pocket and the primary specificity pocket, respectively. (Clearly, in the absence of experimental data, the selection and positioning of the template chemistry becomes a much less secure task, where one would need to consider synthesizing and assaying small representative libraries in order to test the original hypothesis.)

From the models of the template bound to the receptor, it is possible to derive pharmacophores describing the features desired at each substitution position. These may be as tightly or loosely defined as desired, depending on the diversity which is required. For example, in the example of the A-Pro-B library, a 2-point pharmacophore was defined at the B position, containing one aliphatic amine for amide bond formation to the proline C terminus together with a second hydrogen bond donor moiety to make one of several hydrogen bonds accessible within the cation-binding pocket (as this is known to be a key feature of substrate binding), with a loose distance range (5–8 Å) in order to access a diversity of chemistries.

These pharmacophores are used to search a suitable reagent database, in this case the Available Chemicals Directory (ACD), available with the MDL ISIS software (MDL Information Systems Inc., San Leandro, CA, USA). Conformationally flexible 3D searches (using standard ISIS tools) yielded a list of acids which fitted the A position pharmacophore and a list of amines for the B position. Following this, the reagent lists are processed to incorporate any chemical modifications that may be considered by a medicinal chemist, such as removal of a protecting group or performing a facile functional group transformation. This is a useful tool because this is a route to accessing greater diversity than is available in the original reagent database. This process is handled computationally (using our in-house software) by the modeller defining the transformation as a generalised rule in a text-based chemical notation system (similar to SMILES notation), with the result that a particular functional group is automatically identified each time it occurs and is appropriately transformed.

The next phase is the screening of these lists of available reagents to determine quality of binding to the receptor. A hierarchy of screens is applied in order to reduce the list as efficiently as possible, beginning with simple screens such as molecular weight, calculable physical chemistry properties (e.g., log *P*) and undesired 2D substructures. Remaining reagents then are screened against the receptor model. Again, a rulebase (in SMILES-like format) allows the software to determine automatically the way to attach each reagent to the scaffold chemistry (i.e., which bonds to break and form, and which substitution position to use on the scaffold). Functional groups on the reagent (lipophilic sites and hydrogen bonding sites) are automatically flagged, and these interaction sites are used to dock the reagent on to the receptor design model. The objective is to determine whether a particular reagent is capable of being attached to the scaffold while forming good binding interactions with the targeted binding site. Reagents which cannot bind satisfactorily are rejected, e.g., a desired H bond may not be achievable without steric clashing with the receptor. For acceptable substituents, multiple binding conformations are generated and scored (using an empirical scoring function which is an estimation of free energy of binding, together with an estimate of ligand strain).

At the end of this stage, all substituents have been screened and scored, treating each substitution position in isolation. The modeller or medicinal chemist can survey the results to select a subset of substituents for further evaluation, focusing on both high scoring substituents and those with interesting structural diversity (graphical analysis and clustering of the hit-list and scores are useful here). Lists of preferred substituents at each substitution position may then be enumerated combinatorially, and this virtual library is then screened further by allowing a more complete geometry refinement in the receptor model. A subset of designs is then prioritized for synthesis, again by consideration of diversity and predicted affinity, but also taking into account synthetic considerations and other predicted physical or biological properties.

10.3.3 Results

The efficacy of the screening process is exemplified in Table 10.4. A simple query of the ACD for any amine or acid yields lists of many thousands of reagents. Even applying the loose 3D pharmacophore query yields lists of 500 and 900 reagents. At this stage, a medicinal chemist would generally consider clustering these lists on the basis of similarity, in order to pick out a representative sample of reagents. However, this approach is inefficient because many of these reagents are not capable of binding satisfactorily to the receptor. The PRO_SELECT receptor-based screen reduces these lists to about 150 substituents at each position, these being the only substituents for which an acceptable binding mode can be found. This demonstrates that the loose 3D ISIS query retains many false positive hits, and the PRO_SELECT receptor screen is in effect an extremely detailed 3D screen, in which the steric constraints of the binding site are defined fully. The final screens based on predicted affinity and strain energy are important for excluding substituents which bind only poorly. The final lists of substituents can then be clustered and inspected graphically, in order to choose a set of representative synthetic candidates. In the end, only 24 compounds were synthesized from these lists, 18 of which had a binding affinity to thrombin of better than 10 μM, with the best

Table 10.4 An example of the results of substituent screening within PRO_SELECT. The number of substituents retained after each screen is given for the two substitution positions for the A-Pro-B library. The final column indicates the maximum size of library which could be obtained from the lists (i.e., the product of the numbers at each substitution position). The 2D ACD screen was defined as any carboxylic acid or acid chloride at the A position, and any aliphatic amine or hydrazine at the B position. The 3D ACD screen employed simple 2 or 3 point pharmacophores described in the text. See Murray *et al.*, 1997; Young *et al.*, 1998

	No. of accepted substitutents		*No. of compounds in virtual library*
	A position	*B position*	
After 2D ISIS ACD screen	8 803	4 262	37 518 386
After 3D ISIS ACD screen	437	894	390 678
After receptor screen	145	144	20 880
After binding affinity filter	81	65	5 265
After strain energy filter	71	53	3 763
Selected synthesis candidates	8	9	72

at 36 nM. More importantly, a diverse range of substituents was identified (some not used before for thrombin inhibitors) offering markedly improved selectivity over trypsin, and potentially improved pharmacokinetic properties.

Although PRO_SELECT was envisaged originally as a method for ensuring synthetic accessibility in a large library of designs, it may be viewed also as a tool for the design of combinatorial chemistry libraries. The technology exists now to generate and assay huge libraries of combinatorial chemistry compounds, but there is also a prevailing view that there are benefits to focusing or targeting the properties of the library towards the requirements of a given biological receptor. Even large libraries of tens or hundreds of thousands of compounds frequently contain limited or inappropriate diversity, and may have a low hit-rate against a given receptor (though they may well prove very useful against a battery of receptors). Whereas library design traditionally has focused on diversity analysis of the reagents, PRO_SELECT uses the 3D structure of the receptor to screen for substituents which are predicted to bind well. In effect the designed library does not represent all available reagent diversity, but only a very narrow slice of that diversity: as a rule, the receptor structure imposes strict constraints on ligands such that only a narrow range of properties (e.g., size, shape) is tolerated. However, within that tolerated window, PRO_SELECT can explore more exhaustively a diversity of acceptable chemistries. Thus PRO_SELECT would identify all chemistries compatible with a binding pocket, whereas a traditional clustering approach would perhaps generate only a small number of representative examples. Hence PRO_SELECT may well prove of use in designing libraries for combinatorial chemistry applications. Although the above example was for a peptidic inhibitor, the software is flexible enough to cope with any general organic reaction scheme, including examples where the scaffold itself is created during the reaction (e.g., benzodiazepine libraries). Clearly, most applications of combinatorial chemistry currently are directed towards targets for which the receptor structure is not known, and hence PRO_SELECT will be of value only in those projects where a good structural basis is available. It will remain to be seen how far such methods of targeting substituent selection can be developed, so that they can usefully be exploited for examples of indirect drug design, such as when guided only by a QSAR model.

10.4 CONCLUSIONS

This chapter has described the principles behind *de novo* design software, and reviewed its benefits and limitations within an integrated drug discovery programme. *De novo* design is a useful tool for analysing the properties of a binding pocket and for suggesting the type of chemistries that should be pursued, particularly in those cases where novelty and diversity of design are considered important. We have developed several software packages in-house to address various aspects of SBDD. Our motivation for developing

PRO_SELECT was in part to access larger, readily synthesizable *de novo* libraries, but also to offer a novel approach for the more general problem of how best to focus the design of combinatorial chemistry libraries. As such PRO_SELECT has succeeded in providing a pragmatic solution to our requirements for library design in our in-house drug discovery projects.

ACKNOWLEDGEMENTS

The work described in this chapter is the result of the contributions of many scientists at Proteus over the years, including Jin Li, Chris Murray, David Clark, David Westhead and Richard Sykes (computational chemistry), Stephen Young and Phillip Morgan (synthetic chemistry) and Harry Martin (biochemistry).

REFERENCES

Babine, R.E., Bleckman, T.M., Kissinger, C.R., Showalter, R., Pelletier, L.A., Lewis, C., Tucker, K., Moomaw, E., Parge, H.E. and Villafranca, J.E. (1995) Design, synthesis and x-ray crystallographic studies of novel FKBP-12 ligands, *Bioorganic and Medicinal Chemistry Letters*, **5**, 1719–24.

Bohacek, R.S., McMartin, C. and Guida, W.C. (1996) The art and practice of structure-based drug design, *Medicinal Research Reviews*, **16**, 3–50.

Böhm, H.J. (1992) LUDI: rule-based automatic design of new substituents for enzyme inhibitor leads, *Journal of Computer-Aided Molecular Design*, **6**, 593–606.

Böhm, H.J. (1993) Ligand design, in *3D QSAR in Drug Design*, (ed. H. Kubinyi), ESCOM, Leiden, pp. 386–405.

Böhm, H.J. (1994) The development of a simple empirical scoring function to estimate the binding constant for a protein–ligand complex of known three-dimensional structure, *Journal of Computer-Aided Molecular Design*, **8**, 243–56.

Böhm, H.J. and Klebe, G. (1996) What can we learn from molecular recognition in protein–ligand complexes for the design of new drugs?, *Angewandte Chemie International Edition in English*, **35**, 2588–614.

Caflisch, A., Miranker, A. and Karplus, M. (1993) Multiple copy simultaneous search and construction of ligands in binding sites: application to inhibitors of HIV-1 aspartic proteinase, *Journal of Medicinal Chemistry*, **36**, 2142–67.

Clark, D.E., Frenkel, D., Levy, S.A., Li, J., Murray, C.W., Robson, B., Waszkowycz, B. and Westhead, D.R. (1995) PRO_LIGAND: An approach to *de novo* molecular design. 1. Application to the design of organic molecules, *Journal of Computer-Aided Molecular Design*, **9**, 13–32.

Frenkel, D., Clark, D.E., Li, J., Murray, C.W., Robson, B., Waszkowycz, B. and Westhead, D.R. (1995) PRO_LIGAND: an approach to *de novo* molecular design. 4. Application to the design of peptides, *Journal of Computer-Aided Molecular Design*, **9**, 213–25.

Gallop, M.A., Barrett, R.W., Dower, W.J., Fodor, S.P.A. and Gordon, E.M. (1994) Applications of combinatorial technologies to drug discovery. 1. Background and peptide combinatorial libraries, *Journal of Medicinal Chemistry*, **37**, 1233–51.

Gehlhaar, D.K., Moerder, K.E., Zichi, D., Sherman, C.J., Ogden, R.C. and Freer, S.T. (1995) *De novo* design of enzyme inhibitors by Monte Carlo ligand generation, *Journal of Medicinal Chemistry*, **38**, 466–72.

Gillet, V., Johnson, A.P., Mata, P., Sike, S. and Wiliams, P. (1993) SPROUT: a program for structure generation, *Journal of Computer-Aided Molecular Design*, **7**, 127–53.

Gillet, V.J., Myat, G., Zsoldos, Z. and Johnson, A.P. (1995) SPROUT, HIPPO and CAESA: tools for *de novo* structure generation and estimation of synthetic accessibility, *Perspectives in Drug Discovery and Design*, **3**, 34–50.

Goodford, P.J. (1985) A computational procedure for determining energetically favourable binding sites on biologically important macromolecules, *Journal of Medicinal Chemistry*, **28**, 849–57.

Gordon, E.M., Barrett, R.W., Dower, W.J., Fodor, S.P.A. and Gallop, M.A. (1994) Applications of combinatorial technologies to drug discovery. 2. Combinatorial organic synthesis, library screening strategies, and future directions, *Journal of Medicinal Chemstry*, **37**, 1385–401.

Lewis, R.A. and Leach, A.R. (1994) Current methods for site-directed structure generation, *Journal of Computer-Aided Molecular Design*, **8**, 467–75.

Moon, J.B. and Howe, W.J.O. (1991) Computer design of bioactive molecules: a method for receptor-based *de novo* ligand design, *Proteins: Structure, Function, Genetics*, **11**, 314–28.

Murray, C.W., Clark, D.E., Auton, T.R., Firth, M.A., Li, J., Sykes, R.A., Waszkowycz, B., Westhead, D.R. and Young, S.C. (1997) PRO_SELECT: combining structure-based drug design and combinatorial chemistry for rapid lead discovery. 1. Technology, *Journal of Computer Aided-Molecular Design*, **11**, 193–207.

Nishibata, Y. and Itai, A. (1991) Automatic creation of drug candidate structures based on receptor structure: starting point for artificial lead generation, *Tetrahedron*, **47**, 8985–90.

Pisabarro, M.T., Ortiz, A.R., Paolmer, A., Cabre, F., Garcia, L., Wade, R.C., Gago, F., Mauleon, D. and Carganico, G. (1994) Rational modification of human synovial fluid phospholiapse A2 inhibitors, *Journal of Medicinal Chemistry*, **37**, 337.

Rotstein, S.H. and Murcko, M.A. (1993a) GenStar: a method for *de novo* drug design, *Journal of Computer-Aided Molecular Design*, **7**, 123–43.

Rotstein, S.H. and Murcko, M.A. (1993b) GroupBuild: a fragment-based method for *de novo* drug design, *Journal of Medicinal Chemistry*, **36**, 1700–10.

Tschinke, V. and Cohen, N.C. (1993) The NEWLEAD program: a new method for the design of candidate structures from pharmacophoric hypotheses, *Journal of Medicinal Chemistry*, **36**, 3863–70.

Waszkowycz, B., Clark, D.E., Frenkel, D., Li, J., Murray, C.W., Robson, B. and Westhead, D.R. (1994) PRO_LIGAND: an approach to *de novo* molecular design. 2. Design of novel molecules from molecular field analysis (MFA) models and pharmacophores, *Journal of Medicinal Chemistry*, **37**, 3994–4002.

Westhead, D.R., Clark, D.E., Frenkel, D., Li, J., Murray, C.W., Robson, B. and Waszkowycz, B. (1995) PRO_LIGAND: an approach to *de novo* molecular design. 3. A genetic algorithm for structure refinement, *Journal of Computer-Aided Molecular Design*, **9**, 139–48.

Young, S.C., Auton, T.R., Clark, D.E., Li, J., Liebeschuetz, J.W., Lowe, R., Mahler, J., Martin, H., Morgan, P.J., Murray, C.W., Rimmer, A.D., Waszkowycz, B. and Westhead, D.R. (1998) PRO_SELECT: combining structure-based drug design and combinatorial chemistry for rapid lead discovery and optimization. 2. Application to the design of thrombin inhibitors. *Journal of Computer-Aided Molecular Design*. In press.

11

Current and Future Capabilities in Automated Drug Screening and Development

Ian J. Norrington

11.1 INTRODUCTION

The development of systems for automation of procedures in the pharmaceutical research laboratory has been rapid in the last decade, particularly in the area of high throughput screening. However, progress has not just been in terms of throughput. Versatility has been important in bringing a new range of techniques within the scope of automation. Capabilities such as retesting, solid phase extraction, cell permeability studies, redissolution of compounds from libraries, PCR and combinatorial chemistry are now provided by automated systems.

The scope of procedures that can be automated means that technically it is now possible to have a fully automated screening facility. However, such facilities will be feasible only if they are reliable, adaptable and, crucially, economically justifiable. A consideration of the technology and its impact on the drug discovery process, and its productivity, can provide an insight into what the future may bring in this dynamic area.

11.2 GENERAL CONSIDERATIONS

Before considering the specific items of hardware that are now available, it is important to bear in mind the requirements of any automated system. A key requirement is ease of use. In the past, there have been a large number of systems available, some of them quite sophisticated, but requiring a good knowledge of robotics and a very high level of computer literacy on the part

Advances in Drug Discovery Techniques. Edited by Alan L. Harvey © 1998 John Wiley & Sons Ltd.
ISBN 0 471 97509 5

of the operator. In today's environment, and in order for automation to be more widely used, this is no longer acceptable.

A key advance was the development of software that operates in a Microsoft™ Windows™ environment for the implementation and control of procedures. Windows has developed into the common computer currency and the integration of systems within the laboratory means that many people need Windows-based software in order for it to be compatible with their databases. Another important factor is an icon-driven approach. Most people find it easier and more intuitive to design or implement a protocol by using a PC mouse to click on a picture and link it with other functions rather than typing in a string of computer code. For instance, the Tecan workstation software now allows the operator to click on an icon of a tap (representing a plate washer), link it to a torch icon (representing a plate reader) and that is all that is required to set up a wash and read protocol. Obviously, the process can be extended to cover much more demanding applications. These developments over the last five to ten years have dramatically improved the accessibility and ease of use of automated systems.

Robustness is another vital consideration. In the past, analytical systems generally were turned on when they were needed, the particular operation carried out, and then the systems were turned off before moving on to the next piece of equipment. Many facilities are now interested in full automation, in which a system may be left running literally for weeks on end. With the significant investment that a fully automated system may well represent, the ability to gain full utility from the system often will be essential in making the investment economically justifiable. Clearly, the longer the period of time for which an instrument is left unattended, the greater the chance that something may malfunction. This has demanded that equipment must now be made to a higher standard of reliability and robustness.

Speed and throughput are other obvious requirements. Automation does have a number of benefits in high throughput screening, e.g., improved safety, better reproducibility, more productive use of laboratory scientists' time, but it will not be acceptable if it is not faster than working with manual equipment.

Versatility is related to ease of use. The techniques used in pharmaceutical development are continuously evolving, and very few organizations run the same assays and procedures for more than a few years. Often a project will run for about six months, and then the procedures will change, possibly requiring different techniques and equipment. As a result, systems must be modular and be easy to adapt, expand, reconfigure, programme and reprogramme (Palmer, 1996; Reichman and Harris, 1997).

Accuracy and reproducibility are important, even in more routine applications such as primary screening. Accurate and reproducible results will reduce the number of false positive results and the need for retesting, and thus increase the efficiency and throughput of the operation.

Good technical support is required to ensure minimum downtime and provide comprehensive advice and training. With the pressure for higher and higher throughput, it becomes ever more important that systems are up and running all the time.

Many facilities now need to link their systems into a host computer. In many environments a complete data audit is required, so that all elements such as the sample identity, how it was processed and the results generated can be tracked, appended and stored (Lister and Janzen, 1997; Nihara, Kotaki and Nakano, 1997). Instruments operating stand-alone and manual sample tracking and collation of results are now no longer acceptable.

Finally, cost-effectiveness is crucial. The economics of high throughput screening will be addressed after a consideration of some of the specific technical capabilities now available as well as those on the horizon.

11.3 CAPABILITIES IN HIGH THROUGHPUT SCREENING

Techniques such as mother/daughter plate preparation from libraries of compounds are now well established territory for automated systems. However, even here there are developments (see, e.g., Beggs, 1997; Major, 1997). Many people are moving away from conventional 96-well plates to higher density plates for the increased efficiency they offer. Even when 96-well plates are still being used, often two or three compounds, and in some cases, 10 or 20, are being assayed in the same well. After the initial screen, the compounds from the wells testing positive are retested individually.

Automated systems for processing ELISA and fluorometric assays have been available for some time, with the increased sensitivity offered by fluorometric techniques now generally preferred (e.g., Hemmila and Webb, 1997). Ligand binding assays have been a standard technique in pharmaceutical laboratories for several years, but it is only recently that equipment allowing total automation of the technique has been developed (Sweetnam *et al.*, 1993; Sweetnam, Price and Ferkany, 1995).

Scintillation and proximity assays are another developing area which is particularly suited to full automation (see Chapter 3 by Hill). All the components are simply combined in the well, incubated and read, with no phase separation steps. In contrast, ELISA and fluorometric techniques require multiple washing and incubation steps which must be scheduled if more than one reaction plate is to be processed simultaneously. Although scintillation and proximity assays are much simpler to automate, the potential increase in throughput must be balanced against the fact that they are significantly more expensive than the more traditional techniques.

Another technique that has recently moved within the scope of automation is direct cell based assays (see Chapter 4 by McMahon). Other areas include automated dilution protocols where, with some compounds, the activity is enhanced by dilution.

After the initial high throughput screen, often the next step is to retest the positive 'hits'. Automation of this process generally requires linking in with the laboratory mainframe. According to the results of the initial screen, plates must be taken from the storage hotel, and the wells on each plate which tested positive must be identified and selected by some form of robotic processing system for retesting. The results must then be collated with those from the initial screen, secondary 'hits' confirmed, and the next stage of analysis identified. Similarly, samples failing the secondary screen should be identified for any further analysis.

11.4 PREPARATION, FILTRATION AND EXTRACTION

Solid phase extraction has been the subject of considerable research and development by robotic companies over recent years. Automated systems now cater for both negative and positive pressure solid phase extraction techniques, each of which has its particular advantages and disadvantages. Positive pressure has the advantage of allowing total control over flow rates through the column, which can be controlled down to 1 μl per second if necessary.

However, throughput is limited by the need to apply continuous pressure, usually via the pipetting tips of the robotic processor, which restricts the number of samples that can be processed simultaneously to the number of pipetting tips, typically eight for the larger systems.

As negative pressure solid phase extraction is driven by vacuum, the number of samples that can be processed simultaneously is limited only by the size of the vacuum block. If the block holds 100 extraction columns, all 100 can be processed simultaneously. This throughput advantage is offset by variability in flow rates. Although negative pressure generally gives flow rates that are fairly constant at around 90%, there can be differences and problems in individual laboratories. However, generally they function reliably in automated systems as long as it is not critical that columns do not run dry. Racks can be developed to accommodate any kind of column and cater for any combination of throughputs.

Since natural products are an important source for drug discovery programmes, sample clean-up and separation also are major targets for automation. For instance, one area of interest has been the ability to filter out automatically compounds of lower molecular weight, typically below 500 or 600 daltons, which may have pharmaceutical activity. Another growing area is the clean-up of samples generated by combinatorial chemistry, automation of various exchange techniques and purification of DNA samples (Kibbey *et al.*, 1997; Routberg *et al.*, 1997; Van Ausdall and Marshall, 1997).

One specific development has been the introduction of filtration techniques within the format of a 96-well microwell plate. As these have the same dimensions as a standard microtitre plate, they are ideally suited to automated processing using existing robotic systems. This greatly expands the potential throughput as compared with solid phase extraction columns. Typically, the maximum number of solid phase extraction columns that can be handled at one time by a robotic processor would be around 200. With filter plates, the number probably is 10 times greater.

Of particular relevance to drug testing and pharmacokinetics are permeability investigations. Clearly, the passage of drugs across membranes is a key factor in determining their pharmacological characteristics. Generally, the preferred route of administration for a drug is by oral ingestion. It is therefore important to demonstrate that it can cross the epithelium of cells in the gastrointestinal tract and enter the systemic circulation, and so reach its site of action. Less commonly, it may be equally important to show that it does not penetrate into the circulation. Reaction cells have been produced with one-way or two-way membranes to mimic the gut epithelium. Sampling from either side of the membrane allows the proportion of drug that is ingested, and therefore the bioavailability, to be estimated. The results can guide future drug development, influencing such decisions as whether the drug needs to be administered as a prodrug that is more readily taken up and that is converted to the active form once it is in the body. As these processes are kinetic, a clear requirement of any system used to automate them is an ability to sample reliably at specified times. With the scheduling capabilities of modern robotic systems, this is not a particular problem.

11.5 MOLECULAR BIOLOGY

Many procedures in molecular biology, DNA preparation and sequencing for example, are key targets for automation and the focus of much work by robotics companies. The possibilities are too numerous to go into details here, but one example of a system that is in use today provides an indication of potential applications in the future. In this system, a single robotic sample processor is linked with and drives 20 thermal cyclers. This is just one example of what will almost certainly be a growing application of automation to molecular biology.

11.6 COMBINATORIAL CHEMISTRY

Until recently, the main bottleneck in the drug development process was screening. Partly as a result of the advances considered above, the bottleneck now has shifted a step back in the process, to the generation of new compounds. Not surprisingly, automated combinatorial chemistry has now become a focus of activity for many companies, both new and established (Campbell and Wildonger 1997; Cargill and Lebl, 1997; Porco *et al.*, 1997). Some of the considerations for screening apply also to compound generation. It is essential that enough compounds are generated in a relatively short time to justify the outlay on equipment and consumables, particularly for solid phase techniques, which can be expensive. Robustness is particularly important in this area. Many of the chemicals used, TFA for example, are highly corrosive. The development of a system that can cope with these aggressive substances has demanded a lot of time, novel equipment and extensive testing.

Consequently, automated combinatorial chemistry systems are not cheap: a custom developed system can cost more than US$250 000 although off-the-shelf equipment is available from US$150 000. However, they can increase the generation of new compounds 20-fold or more. So, by freeing the bottleneck, automated combinatorial chemistry can justify its initial outlay. At the same time, high yields often are required so that products can be stored in libraries and assessed in an array of tests.

11.7 FULL AUTOMATION

With the range of automated capabilities available, fully automated screening facilities are now feasible (see, e.g., Palmer, 1996; Cook *et al.*, 1997; Eckard *et al.*, 1997; Harding *et al.*, 1997; Moos, 1997; Neary, 1997). The elements needed to make these facilities a reality are currently in existence or in development. Those in development include automated weighing and preparation systems that can weigh out a compound and, on the basis of its chemical structure, make a solution of a defined molarity, as well as automated centrifugation (Ray, 1997). A number of industrial robotic companies are working on systems for fully automated management and transport to and from sample libraries.

Flexibility becomes increasingly important as the extent of automation and the sophistication of the instrumentation increases, and so also do the costs. Fully automated systems must be able to adapt quickly and cost-effectively to changing laboratory requirements.

11.8 COSTS AND RETURNS

Clearly, new technology will be adopted only if it delivers economic benefits. Although specific costs vary for each drug, a typical total cost for bringing a new drug to market is US$200–300 M. Of this total, synthesis or isolation of the compound and screening cost around US$2.5–3 M. Although these stages represent only around 1% of the total investment, they can impact significantly on the economics of the process and the pharmaceutical industry. By getting products through to the pre-clinical stage quicker, automated screening can have positive effects on outward investment and share prices.

In terms of the investment in fully automated systems, the standard standalone robotic sample processing systems typically cost US$15 000–80 000 and give throughputs of 1000–1500 samples per run. The next level up is standalone instruments with internal hotels and analytical stations. These can process around 100 plates at a time, giving throughputs of around 6000 samples per run. These systems also offer the potential for overnight running, integration with other analytical devices and linking into the laboratory mainframe. The level of investment is not substantially higher, but the throughput and capabilities are expanded dramatically.

Fully automated systems cost upwards of US$800 000 with the most sophisticated approaching US$1.8 M. These offer dramatically increased throughput. They can operate six or seven days a week for a 150 hour working week. On this basis, 150 000 samples can be processed each week: equivalent to 7 000 000 samples annually.

Increasing the number of samples that can be screened increases the chance of finding what might, in time, prove to be a successful new drug. Therefore, though the investment may be considerable, the potential returns can make it economically viable.

REFERENCES

Beggs, M. (1997) Use of a two armed Zymark robot for high throughput screening, in *Proceedings of the International Symposium on Laboratory Automation and Robotics 1996*, Zymark Corporation, Hopkington, MA, pp. 206–19.

Campbell, J.B. and Wildonger, R.A. (1997) Managing compound library production by parallel synthesis on automated workstation-based systems, in *Proceedings of the International Symposium on Laboratory Automation and Robotics 1996*, Zymark Corporation, Hopkington, MA, pp. 127–41.

Cargill, J.F. and Lebl, M. (1997) New methods in combinatorial chemistry: robotics and parallel synthesis, *Current Opinions in Chemistry and Biology*, **1**, 67–71.

Cook, J.S., Hynd, B., Kozikowski, B., McOsker, C.C., Paugh, D., Rasmusen, K., Sway, J. and Whitten, J. (1997) Establishing a high throughput screening facility at Procter & Gamble Pharmaceuticals, in *Proceedings of the International Symposium on Laboratory Automation and Robotics 1996*, Zymark Corporation, Hopkington, MA, pp. 197–205.

Eckard, P., Delzer, J., Eming, F., Guhi, S., Janocha, R., Market, C., Paul, G., Seega, J. and Wemet, W. (1997) Integrated drug discovery: thriving on organizational and technological improvements, in *Proceedings of the International Symposium on Laboratory Automation and Robotics 1996*, Zymark Corporation, Hopkington, MA, pp. 233–44.

Harding, D., Banks, M., Fogarty, S. and Binnie, A. (1997) Development of an automated high-throughput screening system: a case history, *Drug Discovery Today*, **2**, 385–90.

Hemmila, I. and Webb, S. (1997) Time-resolved fluorometry: an overview of the labels and core technologies for drug screening applications, *Drug Discovery Today*, **2**, 373–81.

Kibbey, C.E., Robertson, G.A., MacDonald, A. and DeWitt, S.H. (1997) An automated system for the purification of combinatorial libraries by semi-preparative HPLC, in *Proceedings of the International Symposium on Laboratory Automation and Robotics 1996*, Zymark Corporation, Hopkington, MA, pp. 336–48.

Lister, M.D. and Janzen, W. (1997) Management of a centralized high throughput screening facility, in *Proceedings of the International Symposium on Laboratory Automation and Robotics 1996*, Zymark Corporation, Hopkington, MA, pp. 43–50.

Major, J. (1997) Challenges and opportunities in high throughput screening: implications for new technologies, in *Proceedings of the International Symposium on*

Laboratory Automation and Robotics 1996, Zymark Corporation, Hopkington, MA, pp. 25–34.

Moos, W.H. (1997) Drug discovery meets biotech: how automation enables the 'better-faster-cheaper' paradigm, in *Proceedings of the International Symposium on Laboratory Automation and Robotics 1996*, Zymark Corporation, Hopkington, MA, pp. 1–14.

Neary, C. (1997) The present and future of automation in drug discovery, in (ed. W. Hori), International Business Communications, Inc., Southborough, pp. 1.1.1–66.

Nihira, S., Kotaki, H. and Nakano, A. (1997) High throughput screening through the use of robotics and high performance data handling system, in *Proceedings of the International Symposium on Laboratory Automation and Robotics 1996*, Zymark Corporation, Hopkington, MA, pp. 259–68.

Palmer, M.A.J. (1996) High throughput screening on a low budget, *Nature Biotechnology*, **14**, 513–15.

Porco, J.A.J., Deegan, T., Devonport, W., Gooding, O.W., Heisier, K., Labadie, J.W., Newcomb, B., Nguyen, C., van Eikaren, P., Wong, J. and Wright, P. (1997) Automated chemical synthesis: from resins to instruments, *Molecular Diversity*, **2**, 197–206.

Ray, B. (1997) New approaches for material handling at Glaxo Wellcome, in *Proceedings of the International Symposium on Laboratory Automation and Robotics 1996*, Zymark Corporation, Hopkington, MA, pp. 398.

Reichman, M. and Harris, A.L. (1997) Practical high throughput screening, in *Annual Reports in Combinatorial Chemistry and Molecular Diversity* (eds W.H. Moos, M.R. Pavia, B.K. Kay, and A.D. Ellington), ESCOM, Leiden, pp. 273–86.

Routburg, M., Swenson, R., Schmitt, B., Washington, A., Mueller, S, Hochlowski, J., Maslana, G., Minin, B., Matuszak, K., Searle, P. and Pan, J. (1997) Implementation of an automated purification/verification system, in *Proceedings of the International Symposium on Laboratory Automation and Robotics 1996*, Zymark Corporation, Hopkington, MA, pp. 360–74.

Sweetnam, P.M., Caldwell, L., Lancaster, J., Bauer Jr., C., McMillan, B., Kinnier, W.J. and Price, C.H. (1993) The role of receptor binding in drug discovery, *Journal of Natural Products*, **56**, 441–55.

Sweetnam, P.M., Price, C.H. and Ferkany, J.W. (1995) Mass ligand screening as a tool for drug discovery and development, in *Burger's Medicinal Chemistry and Drug Discovery*, 5th Edn, Vol. 1, *Principles and Practice* (ed. M.E. Wolff), Wiley, New York, pp. 697–731.

Van Ausdall, D.A. and Marshall, W.S. (1997) Automated high throughput mass spectral analysis of oligonucleotides, in *Proceedings of the International Symposium on Laboratory Automation and Robotics 1996*, Zymark Corporation, Hopkington, MA, pp. 116-22.

12

Antisense Oligonucleotide Therapeutics

C. Frank Bennett, Nicholas M. Dean and Brett P. Monia

12.1 INTRODUCTION

With the complete sequencing of the human genome well underway, a sense of excitement exists in the biomedical community anticipating the numerous benefits to human health this information will provide. One of the most direct applications of this endeavour is antisense technology, in which the primary sequence information is used to design short oligonucleotides to hybridize to a specific mRNA by Watson–Crick base pairing rules (antisense oligonucleotide). Upon binding to the targeted mRNA in a cell, the antisense oligonucleotide prevents expression of the protein product encoded by the targeted RNA. An oligonucleotide may inhibit target gene expression by several mechanisms of which oligonucleotide-induced degradation of the target RNA by the cellular enzyme RNase is exploited most commonly (Helene and Toulme, 1990; Bennett and Crooke, 1994; Crooke, 1995; Crooke and Bennett, 1996). Antisense oligonucleotides can be used also as pharmacological tools to dissect out roles of specific gene products in complex biological pathways, and potentially as therapeutic agents. There are numerous publications describing the pharmacological activity of antisense oligonucleotides in cellular assays, *in vivo*, and more recently in patients (Crooke, 1992, 1995; Bishop *et al*., 1996; Crooke and Bennett, 1996; Glover *et al*., 1997; Webb *et al*., 1997; Yacyshyn *et al*., 1997a). Conceptually, antisense oligonucleotides represent a truly 'rational' approach for drug discovery. However, in practice, as with most scientific endeavours, successful application of the technology is not as simple as it would at first seem (Stein and Cheng, 1993; Stein and Krieg, 1994; Wagner, 1994; Branch 1996; Crooke and Bennett, 1996; Bennett, 1997). Issues have been raised primarily concerning interaction of the oligonucleotide with

Advances in Drug Discovery Techniques. Edited by Alan L. Harvey
ISBN 0 471 97509 5

non-target molecules, in particular non-nucleic acid molecules. While these interactions can contribute to the pharmacological activity of an oligonucleotide, it is possible to demonstrate specific inhibition of a gene product by an antisense mechanism of action, provided that the oligonucleotides are selected carefully, correct controls are included and complete dose–response curves are performed (Chiang *et al.*, 1991; Monia *et al.*, 1992, 1993, 1996b,c; Bennett, 1997). This chapter discusses applications of antisense technology in providing potential therapeutic agents. The reader is referred to several recent reviews, which discuss more basic concepts of antisense technology and applications as a pharmacological tool (Crooke, 1992, 1995; Bennett and Crooke, 1994; Crooke and Bennett, 1996; Dean *et al.*, 1996a; Sharma and Narayanan, 1995).

12.2 FIRST GENERATION OLIGONUCLEOTIDES

Rapid degradation of natural DNA by serum and cellular nucleases limits their utility as antisense agents *in vivo* (Hoke *et al.*, 1991; Shaw *et al.*, 1991). A number of chemical modifications, which increase the resistance of the oligonucleotide to nuclease degradation, have been described (Milligan, Matteucci and Martin, 1993; Sanghvi and Cook, 1994; Altmann *et al.*, 1996; Crooke *et al.*, 1996; Griffey *et al.*, 1996; Matteucci, 1996; McKay *et al.*, 1996; Monia *et al.*, 1996a). The most commonly used modified oligonucleotides are phosphorothioate oligodeoxynucleotides, in which one of the non-bridging oxygen atoms in the phosphate backbone is replaced with sulfur (De Clercq, Eckstein and Merigan, 1969). This modification dramatically increases resistance to nuclease degradation, maintains the negative charge on the backbone and is capable of supporting RNase H activity. Phosphorothioate oligodeoxynucleotides are synthesized on commercial oligonucleotide synthesizers, the materials required for synthesis are readily available, and there are commercial suppliers of custom oligonucleotides. Although there are clear limitations for phosphorothioate oligodeoxynucleotides (Stein and Krieg, 1994; Stein, 1995, 1996), they represent the most economical approach for developing antisense inhibitors of specific gene products. In fact, 11 out of 12 oligonucleotides that have entered clinical trials are phosphorothioate oligodeoxynucleotides (Table 12.1).

12.2.1 Pharmacokinetics

The plasma pharmacokinetics of phosphorothioate oligodeoxynucleotides has been studied extensively in rodents, non-human primates and humans. Phosphorothioate oligodeoxynucleotides distribute from plasma to tissues with plasma half-lives ranging from 30 min to 1h after intravenous injection (Agrawal, Temsamani and Tang 1991; Cossum *et al.*, 1993; Agrawal *et al.*, 1995a; Zhang *et al.*, 1995b; Geary *et al.*, 1997a,b; Glover *et al.*, 1997). Plasma clearance is biphasic with terminal half-lives reported to range from 40 h to

Table 12.1 Status of antisense oligonucleotides which have entered clinical trials

Drug	*Molecular target*	*Indications*	*Delivery route*	*Company*	*Status*
ISIS 2105	HPV 6 and 11 E2 gene product	Genital warts	Intradermal	Isis Pharmaceuticals	Terminated after Phase II
GEM 91	HIV gag protein	HIV	Intravenous	Hybridon	Terminated in Phase II
ISIS 2922 (Fomivirsen)	CMV immediate early 2 gene	CMV retinitis	Intravitreal	Isis Pharmaceuticals	Phase III
GEM 132	CMV UL36 gene	CMV retinitis	Intravenous, intravitreal	Hybridon	Phase I
OL(1)p53	P53	Haematological malignancy	Intravenous	University Nebraska/ Lynx Therapeutics	Completed Phase I
	C-myb	Chronic myelogenous leukaemia	Intravenous	University of Pennsylvania/Lynx Therapeutics	Completed Phase I
CGP 64128A/ ISIS 3521	Protein kinase C-α	Cancer, misc.	Intraveous	Novartis/Isis Pharmaceuticals	Phase II
CGP 69846A/ ISIS 5132	C-raf kinase	Cancer, misc.	Intravenous	Novartis/Isis Pharmaceuticals	Phase II
Genta 3139	Bcl-2	Non Hodgkins lymphoma	Subcutaneous	Genta	Completed Phase I
ISIS 2503	Ha-ras	Cancer, misc.	Intravenous	Isis Pharmaceuticals	Phase I
ISIS 2302	ICAM-1	Crohn's disease, ulcerative colitis, rheumatoid arthritis, psoriasis, renal transplant	Intravenous, Subcutaneous	Isis Pharmaceuticals/ Boehringer Ingelheim	Phase II
LR3280	c-myc	Restenosis	Topical (intra-arterial)	Lynx Therapeutics	Phase II

60 h. At doses below 20 mg/kg, the bulk of oligonucleotide appears to be bound to plasma proteins (Cossum *et al.*, 1993; Sands *et al.*, 1994; Kumar, Tewary and Iversen, 1995; Crooke *et al.*, 1996). Higher doses of oligodeoxynucleotides may saturate protein binding, resulting in the appearance of free drug in the urine (Agrawal, Temsamani and Tang, 1991). Several studies have reported nonlinear changes in area under the curve (AUC) and volume of distribution with respect to dose at lower dose ranges, suggesting that also there may be a saturable component to the disposition of oligonucleotides (Geary *et al.*, 1997b; Glover *et al.*, 1997; Levin *et al.*, 1997, Qian *et al.*, 1997). Phosphorothioate oligodeoxynucleotides exhibit good bioavailability following subcutaneous and intradermal dosing (Cossum *et al.*, 1994; Agrawal *et al.*, 1995a; Geary *et al.*, 1997a; Levin *et al.*, 1997). Oral bioavailability of phosphorothioate oligodeoxynucleotides is low in normal bowel, without the aid of a formulation to enhance oral absorption. One of the factors which limit oral bioavailability is degradation of the oligonucleotide by nucleases present in intestinal contents (Agrawal *et al.*, 1995b).

Phosphorothioate oligodeoxynucleotides distribute to most peripheral tissues with highest concentrations of drug found in kidney and liver. Other organs which accumulate significant levels of oligonucleotide include spleen, bone marrow and skin. Recent studies have examined the cellular distribution of oligonucleotides within organs, with kidney being best characterized (Cossum *et al.*, 1993; Sands *et al.*, 1994; Plenat *et al.*, 1995; Rappaport *et al.*, 1995; Butler, Stecker and Bennett, 1997). Autoradiographic studies and dissection of kidney after administration of radiolabelled oligonucleotide suggest that the highest concentrations occur in the cortex of the kidney. This observation was confirmed using higher resolution autoradiography, administration of fluorescent oligonucleotide or immunohistochemistry in which phosphorothioate oligodeoxynucleotides were localized to cytoplasmic vesicles in the epithelial cells of the proximal convoluted tubules in the kidney. Lesser amounts of oligonucleotide accumulate in the glomerulus and distal tubules. In liver the bulk of oligonucleotide accumulates in Kupffer cells and sinusoidal endothelial cells (Plenat *et al.*, 1995; Bijsterbosch *et al.*, 1997; Butler, Stecker and Bennett, 1997). Approximately a 10-fold lower concentration of drug was found in hepatocytes. As the dose of oligonucleotide is increased, accumulation of oligodeoxynucleotide in hepatocytes increases, while oligodeoxynucleotide concentrations in Kupffer cells and sinusoidal endothelial cells tend to saturate (Bijsterbosch *et al.*, 1997; Graham *et al.*, 1998). Studies suggest that accumulation of oligonucleotides in sinusoidal endothelial cells and kidney epithelial cells may be mediated in part by binding to scavenger receptors (Sawai *et al.*, 1995; Bijsterbosch *et al.*, 1997). In other tissues, phosphorothioate oligodeoxynucleotides can be found in multiple cell types, but macrophages and cells resembling fibroblasts tend to accumulate the highest levels of oligonucleotide (Butler, Stecker and Bennett, 1997). It remains to be demonstrated whether there is a direct correlation between sensitivity to antisense effects and bulk accumulation of oligonucleotide in the different cell types in the tissues.

Although phosphorothioate oligodeoxynucleotides exhibit increased resistance to serum and cellular nucleases, the compounds are metabolized after administration (Agrawal, Temsamani and Tang, 1991; Cossum *et al.*, 1993; Agrawal *et al.*, 1995a; Crooke *et al.*, 1996; Geary *et al.*, 1997a). The predominant route of oligonucleotide degradation appears to be 3′-exonuclease activity; however, there is evidence that 5′-exonuclease and endonuclease also will degrade phosphorothioate oligodeoxynucleotides. As early as 10 min after intravenous injection, chain shortened metabolites can be isolated from plasma (Geary *et al.*, 1997b). The rate of loss of parent oligonucleotide tends to plateau after 30 min, which may be attributable to the differential nuclease sensitivity of the Rp and Sp phosphorothioate linkages at the 3′ end of the oligonucleotides, with the Rp stereoisomer being more nuclease sensitive (Koziolkiewicz *et al.*, 1997). Degradation of phosphorothioate oligodeoxynucleotides also can be detected in tissues with tissue half-lives ranging from 24 h to greater than 120 h (Geary *et al.*, 1997b; Levin *et al.*, 1997). Clearance from monkey kidney cortex is slower than from other tissues; therefore, it would be anticipated that, with repeat dosage, kidney cortex may accumulate more drug than other tissues (Levin *et al.*, 1997).

12.2.2 Pharmacology

Phosphorothioate oligodeoxynucleotides have been used to inhibit the expression of a wide variety of gene products with applications in most subdisciplines of pharmacology, including virology, oncology, cardiovascular, pulmonary pharmacology, immunopharmacology, and neuropharmacology (Crooke and Bennett, 1996). In some systems, such as anti-viral assays, it is difficult to distinguish between the desired antisense effects of the oligonucleotide and the undesired non-antisense effects (Stein and Cheng, 1993; Wagner, 1994; Crooke and Bennett, 1996). However, in other systems such as neuropharmacology, oncology and immunopharmacology, data would support that the observed pharmacological effects of the oligonucleotides are due to specific antisense effects. The key to the successful use of phosphorothioate oligodeoxynucleotides is the identification of potent inhibitors, which can be accomplished by screening multiple oligonucleotides in assays which directly measure the target gene of interest rather than downstream events (Chiang *et al.*, 1991; Barbour and Dennis, 1993; Coliege *et al.*, 1993; Bennett *et al.*, 1994; Dean and McKay, 1994; Dean *et al.*, 1994; Stepkowski *et al.*, 1994; Monia *et al.*, 1996b). In most cases (but not all), cationic lipids or other means of intracellular delivery of the oligonucleotide were used to deliver the antisense oligonucleotide to cultured cells (Chiang *et al.*, 1991; Bennett *et al.*, 1992). Cationic lipids or other delivery means do not appear to be a prerequisite for *in vivo* experiments, suggesting that *in vitro* cell culture experiments do not predict the *in vivo* pharmacokinetics of phosphorothioate oligodeoxynucleotides. As several recent reviews have summarized the pharmacological effects of antisense oligonucleotides, we will highlight two examples of viral systems in which the effects of the oligonucleotides probably are due to a combination of specific antisense effects and non-specific effects, and three examples in which the data would support that the oligonucleotides are working by an antisense mechanism of action.

(a) Use of antisense oligonucleotides as antiviral therapy

GEM 91 is a 25-mer phosphorothioate oligodeoxynucleotide designed to hybridize to a conserved region of the *gag* region of HIV RNA (Agrawal and Tang, 1992). GEM 91 inhibits viral replication in short term viral assays in a concentration dependent manner, while a 4–5-fold higher concentration of a random mixture of 25-mer phosphorothioate oligodeoxynucleotides (complexity = 4^{25} unique molecules) was required to inhibit viral replication to a similar extent (Lisziewicz *et al.*, 1994). Other studies have demonstrated that acute HIV viral assays are particularly sensitive to the non-antisense effect of phosphorothioate oligodeoxynucleotides (Agrawal *et al.*, 1989; Stein *et al.*, 1991; Lisziewicz *et al.*, 1993). In chronic HIV assays, GEM91 suppressed viral replication for greater than 30 days, while the random mixture of oligodeoxynucleotides suppressed viral replication for only 10 days. GEM 91 was found to be effective against several viral isolates in primary lymphocytes and macrophages and exhibited selectivity in comparison with the random mixture. Since a random mixture of 4^{25} sequences was used as a control, it is difficult to conclude that GEM 91 inhibits viral replication in a sequence specific manner. Based upon these data, it is likely that at least part of the antiviral activity exhibited by GEM 91 is due to a non-antisense effect.

Screening a series of phosphorothioate oligodeoxynucleotides targeting human cytomegalovirus (HCMV) DNA polymerase gene, or RNA transcripts of the major immediate-early regions 1 and 2 (IE1 and IE2), resulted in the identification of ISIS 2922 as a potent inhibitor of HCMV infection of human dermal fibroblasts (Azad *et al.*, 1993). ISIS 2922 is a 21-mer phosphorothioate oligodeoxynucleotide targeting the coding region of the immediate-early 2 gene. ISIS 2922 inhibits viral protein expression, as measured by an ELISA detecting a HCMV late protein product, in fibroblasts with EC_{50} value of 0.1 μM. Non-complementary phosphorothioate oligodeoxynucleotides exhibit an EC_{50} value of 2 μM, 20-fold higher than ISIS 2922. In a plaque reduction assay, ISIS 2922 exhibited an IC_{70} value of 0.1 μM, while a control oligonucleotide exhibited an IC_{70} value of 2 μM. These data suggest that HCMV infection of human dermal fibroblast can be inhibited non-specifically by higher concentrations of phosphorothioate oligodeoxynucleotides; however, ISIS 2922 is approximately 20-fold more effective than non-specific oligonucleotides. ISIS 2922 reduced IE1 and IE2 proteins in infected cells, as did control oligonucleotides at 10-fold higher concentrations. Since the IE1 and IE2 gene products arise from a common pre-mRNA, these results suggest that the oligonucleotide hybridizes to the pre-mRNA. Deletion of sequences from the 5′ and/or 3′ end of the oligonucleotides reduced antiviral activity, while introduction of mismatches in the interior of the oligonucleotide did not reduce antiviral activity significantly, although this did reduce hybridization to the target RNA. These data suggest that the antiviral activity of ISIS 2922 may not be due entirely to an antisense effect. To address this issue in more detail, U373 cells permanently transfected with the IE72 or IE55 polypeptides (derived from the IE1 and IE2 genes, respectively) were treated with ISIS 2922 (Anderson *et al.*, 1996). ISIS 2922 reduced IE55 but not IE72 protein and RNA levels in a sequence specific manner suggesting that reduction of IE55 expression occurs by an RNase H dependent mechanism. As the construct used to express IE72 protein does not contain the ISIS 2922 binding site, these data would support that ISIS 2922 reduces IE55 expression by an antisense mechanism of action. The antiviral activity of ISIS 2922 was not due to immune stimulation by the CpG motifs in the oligonucleotide (Krieg *et al.*, 1995), since methylation of all of the cytosines or only 2 cytosines in the CpG motifs did not reduce antiviral activity. These studies in aggregate suggest that ISIS 2922 is a potent inhibitor of CMV replication which is capable of inhibiting viral gene expression by an antisense mechanism of action, but also may inhibit viral replication by a non-antisense mechanism of action at higher concentrations. Whether both mechanisms of action are operational in the clinic remains to be elucidated.

(b) Use of antisense oligonucleotides for cancer therapy

Protein kinase C (PKC) was identified originally as a serine/threonine kinase involved in mediating intracellular responses to a variety of growth factors, hormones and neurotransmitters (Nishizuka, 1992). Molecular cloning studies have revealed that PKC exists as a family of at least 11 closely related isozymes, which are sub-divided on the basis of certain structural and biochemical similarities (Asaoka *et al.*, 1992; Nishizuka, 1992; Dekker and Parker, 1994). Considerable experimental evidence exists for a role for PKC in some abnormal cellular process, such as inflammation, tumour promotion and carcinogenesis. We have developed antisense oligonucleotides to target individual members of the PKC family, both as research tools and as potential drugs (Dean and McKay, 1994; Dean *et al.*, 1994; Levesque *et al.*, 1996; McKay *et al.*, 1996; Liao *et al.*, 1997). Antisense oligonucleotides which specifically inhibit expression of PKC-α either in mouse or human cell-lines have been used to identify cellular processes which are governed by this PKC isozyme. Initial studies examined the role PKC-α plays in phorbol ester mediated transcriptional regulation of the cell adhesion molecule ICAM-1 in human lung carcinoma cells A549 (Dean *et al.*, 1994). Although down regulation of PKC-α

with specific antisense oligonucleotides attenuated phorbol ester induced ICAM-1 expression, the inhibition was not complete, suggesting that in addition to PKC-α, other PKC isozymes may be playing a role. PKC has been suggested to play a role also in receptor regulation, by down-regulating responses subsequent to receptor activation. Recent studies have found that PKC-α inhibition will modify this process. In particular, reducing PKC-α expression with antisense oligonucleotides was found to attenuate phorbol ester mediated reduction of bradykinin induced calcium mobilization in A549 cells (Levesque *et al.*, 1996).

To understand better the utility of oligonucleotide inhibitors of PKC expression *in vivo*, we have used oligonucleotides targeting murine PKC-α to inhibit expression of PKC-α in mice. A phosphorothioate oligodeoxynucleotide, ISIS 4189, was identified which inhibited PKC-α expression in mouse cell lines. ISIS 4189, formulated in saline, promoted a dose-dependent reduction in PKC-α mRNA expression in liver, with an ID_{50} of about 20mg/kg (Dean and McKay, 1994). The oligonucleotide failed to consistently reduce PKC-α mRNA expression in other organs; however, sometimes decreases in kidney PKC-α were apparent. Scrambled control oligonucleotides failed to reduce PKC-α mRNA levels in liver, suggesting that ISIS 4189 is working by an antisense mechanism of action. Demonstrating that the oligonucleotide did not effect expression of other isozymes provided further support that it was inhibiting PKC-α by an antisense mechanism. These results are consistent with the reported pharmacokinetics of systemically administered phosphorothioate oligodeoxynucleotides, in that a primary organ of deposition is the liver (Agrawal, Temsamani and Tang 1991; Cossum *et al.*, 1993). These results demonstrated for the first time the utility of phosphorothioate oligodeoxynucleotides as inhibitors of 'host' gene expression after systemic administration.

The effects of the human specific PKC-α phosphorothioate oligodeoxynucleotide ISIS 3521 have been examined on the growth of human tumour xenografts in nude mice. In one study, the oligodeoxynucleotide was found to inhibit the growth of three different subcutaneously grown human tumour cell lines: T-24, a bladder carcinoma, A549, a non-small-cell lung carcinoma, and Colo 205, a colon carcinoma (Dean *et al.*, 1996b). ISIS 3521/CGP 64128A was found to inhibit the growth of all three tumour xenografts in a dose-dependent manner, with ID_{50} values for the growth inhibition of between 0.06 mg/kg and 0.6 mg/kg as daily intravenous injections. Three control phosphorothioate oligodeoxynucleotides were without effect on the growth of the tumours at doses as high as 6.0 mg/kg. Analysis of PKC-α expression in the tumour tissue by immunohistochemistry revealed positive staining present in the cytoplasm and occasionally in the nuclei of tumour cells in animals treated with either saline or a scrambled control phosphorothioate oligodeoxynucleotide. By contrast, tumours treated with ISIS 3521/CGP 64128A showed reduced staining for PKC-α.

In a second series of independent studies, ISIS 3521/CGP 64128A was used to suppress the growth of U-87 glioblastoma tumour cells in nude mice (Yazaki *et al*, 1996). This cell line was chosen for study as it has been shown previously to be sensitive to growth inhibition by transfection with an antisense PKC-α cDNA. ISIS 3521/CGP 64128A reduced the growth of these tumour cells when implanted both subcutaneously and intracranially, whereas the scrambled control compound was without effect. This resulted in a doubling in median survival time of the animals with intracranially implanted tumours, with 40% of the treated animals being long term survivors. Levels of ISIS 3521/CGP 64128A and the scrambled control oligodeoxynucleotide within tumour tissue were determined by capillary gel electrophoresis and both were found to be present at about 2 μM after 21 daily i.p. doses of 20 mg/kg oligodeoxynucleotide. ISIS 3521/CGP 64128A also reduced the expression of PKC-α in the tumour tissue, but not that of PKC-ε or PKC-ζ.

These studies demonstrate two important points. First, phosphorothioate oligodeoxynucleotides targeting murine PKC-α can be used selectively to suppress PKC-α expression in mice after systemic administration, in the absence of any lipid formulation. Second, an oligodeoxynucleotide targeting human PKC-α can inhibit the growth of human tumour xenografts selectively in nude mice by a mechanism entirely consistent with an antisense mechanism of action.

The *raf* family of gene products also encodes for serine/threonine-specific protein kinases which play a pivotal role in mitogenic signaling events (Rapp, 1991; Howe *et al.*, 1992; Williams, Roberts and Li, 1992; Daum *et al.*, 1994). There are three known isozymes of *raf* kinase, A-*raf* B-*raf* and C-*raf*. C-*raf* kinase associates with *ras* and transmits signals downstream of *ras* genes in the MAP kinase pathway. In addition, C-*raf* kinase has been shown to associate with *bcl*-2 and also may play a function in regulating apoptosis (Wang *et al.*, 1994). These data suggest that inhibitors of C-*raf* kinase may be of value in regulating abnormal cell proliferation such as cancer.

To identify antisense oligodeoxynucleotides capable of inhibiting human C-*raf* kinase gene expression, a series of phosphorothioate oligodeoxynucleotides were designed and tested for inhibition of C-*raf* mRNA levels in A549 lung carcinoma cells (Monia *et al.*, 1996b). Oligodeoxynucleotides used in this analysis targeted various regions of the C-*raf* mRNA including the 5′ untranslated region (UTR), translation initiation AUG, the coding region and the 3′ UTR, and all were 20 bases in length. Other groups have reported on the activities of antisense-designed oligodeoxynucleotides targeted to the translation initiation AUG of human C-*raf* kinase (Kasid *et al.*, 1989; Soldatenkov *et al.*, 1997). Reductions in C-*raf* mRNA levels were observed following treatment with only a small subset of the oligodeoxynucleotides targeting various regions of the C-*raf* kinase mRNA (Monia *et al.*, 1996c). Furthermore, for those oligonucleotides which did cause reduced C-*raf* mRNA levels, the degree of activity varied greatly. The most potent antisense inhibitor identified from this screen was ISIS 5132 (CGP 69846A), which targets the 3′UTR of the C-*raf* message.

The sequence requirements for inhibiting C-*raf* mRNA and protein expression have been examined thoroughly *in vitro* by comparing the dose-dependent effects of ISIS 5132 with a series of 'mismatched oligonucleotides' containing between one and seven mismatches within the ISIS 5132 sequence (Monia *et al.*, 1996c). Melting temperatures (T_m) along with corresponding dissociation constants (K_d) were determined under cell-free conditions for the duplexes of each of these oligodeoxynucleotides with a 20-mer oligoribonucleotide (RNA) complementary to ISIS 5132. As expected for Watson–Crick-based hybridization, affinity decreased as the number of mismatches contained within the ISIS 5132 sequence increased. No cooperative binding was observed for oligonucleotides containing more than six mismatches. The IC_{50} for ISIS 5132-mediated reduction of C-*raf* mRNA levels in A549 tumour cells in culture is approximately 100 nM (Monia *et al.*, 1996b). As expected, none of the mismatched oligonucleotides was as potent as ISIS 5132 in inhibiting C-*raf* mRNA expression. Furthermore, inhibition of C-*raf* mRNA levels diminished gradually with an increase in the number of mismatches within the ISIS 5132 sequence. Incorporation of a single mismatch resulted in a twofold loss in potency. No activity was observed for oligodeoxynucleotides containing more than four mismatches. These findings are predicted results if the effects of ISIS 5132 on C-*raf* mRNA expression were occurring through a mechanism based on Watson–Crick hybridization to cellular C-*raf* RNA.

In addition to sequence specificity, the effects of ISIS 5132 on other targets were examined to demonstrate target specificity. ISIS 5132 failed to inhibit the expression of the structurally and functionally related A-*raf* kinase and B-*raf* kinase isozymes; also ISIS 5132 did not inhibit the expression of the housekeeping gene for glyceraldehyde 3-phosphate dehydrogenase (Cioffi *et al.*, 1996; Monia *et al.*, 1996b,c). In separate studies, we have identified antisense inhibitors targeted to human A-*raf* and human B-*raf* employing a similar approach to that described for the identification of ISIS 5132. The most active A-*raf* antisense inhibitor identified was ISIS 9069, which targets the translational stop codon of human A-*raf* mRNA (Cioffi *et al.*, 1996). The most active B-*raf* antisense inhibitor discovered was ISIS 13741, which targets the coding region of human B-*raf* mRNA. Employing studies that are similar to those described above for ISIS 5132, we have found that these oligonucleotides are highly sequence specific for their targeted isozymes.

We have studied the effects of antisense inhibitors targeted to C-*raf* kinase on downstream signalling events as well as on cellular proliferation (Cioffi *et al.*, 1996; Monia *et al.*, 1996b, Schulte *et al.*, 1996). Inhibiting the expression of a single *raf* kinase isozyme can abrogate the MAP kinase phosphorylation cascade in response to specific growth factors and cytokines. C-*raf* protein levels were reduced by treating serum-starved A549 cells with ISIS

5132 for 48 h, after which the cells were stimulated with epidermal growth factor (EGF) or phorbol ester followed by quantitative measurements of MAP/ERK kinase activity. Reduction in C-*raf* protein levels by ISIS 5132 almost completely inhibited stimulation of MAP kinase activity by EGF, but had no effect on MAP kinase stimulation by phorbol ester (Monia, 1997). The mismatched control ODN had no effect on MAP kinase stimulation by either agent. These results are consistent with a direct role of C-*raf* in mediating MAP kinase stimulation in response to EGF and demonstrate that activation of MAP kinase by phorbol ester, which requires the activity of protein kinase C (Nishizuka, 1992), occurs independently of C-*raf* in A549 cells. However, we have shown that C-*raf* antisense inhibition can block phorbol ester mediated stimulation of MAP kinase in mouse 3T3 cells, demonstrating the cell-specific nature of cell signalling mechanisms (Schulte *et al.*, 1996). These results demonstrate also that inhibition of a single *raf* kinase isozyme in the MAP kinase pathway is capable of abrogating MAP kinase stimulation almost completely despite the fact that levels of other *raf* kinase family members (A-*raf* and B-*raf*) are unchanged. This finding suggests that, at least in A549 cells, A-*raf* and B-*raf* function differently relative to C-*raf* within the MAP kinase pathway, and indicates that redundancy of function within the *raf* kinase gene family does not exist sufficiently to mediate MAP kinase stimulation in response to certain agents in the absence of C-*raf*. Studies are in progress examining the effects of A-*raf* and B-*raf* antisense inhibitors on MAP kinase stimulation by various agents and in additional cell lines.

ISIS 5132 is a potent anti-proliferative agent against a variety of tumour types when administered intravenously, in saline, at daily doses of 0.01–10 mg/kg (Altmann *et al.*, 1996; Monia *et al.*, 1996b,c; Geiger *et al.*, 1997). Administration of ISIS 5132 to mice bearing subcutaneously implanted A549 tumours resulted in a time-dependent reduction in C-*raf* mRNA levels in tumours as determined by Northern blot analysis. However, no effects on C-*raf* mRNA levels were observed following administration of a control ODN. The antitumour effects of ISIS 5132 are highly sequence specific and the rank order antitumour potency of mismatched analogues is predicted from the Watson–Crick hybridization mechanism. Even the incorporation of a single mismatch resulted in a significant loss of antitimour activity. No antitumour activity was observed for phosphorothioate oligodeoxynucleotides containing more than two mismatches. These results, along with the effects observed on C-*raf* mRNA levels in tumours *in vivo*, strongly support an antisense mechanism underlying the *in vivo* antitumour properties of ISIS 5132. Combinations of ISIS 5132 with conventional chemotherapeutic agents demonstrated that the combinations were additive or supra-additive for several combinations (Geiger *et al.*, 1997). In no instance did ISIS 5132 antagonize the effect of the chemotherapeutic agent.

(c) Use of antisense oligonucleotides for the treatment of inflammatory diseases

In addition to targeting gene products implicated in viral replication or cancer, antisense oligonucleotides have been used to inhibit the expression of gene products, and this may have utility for the treatment of inflammatory diseases. Intercellular adhesion molecule 1 (ICAM-1) is a member of the immunoglobulin gene family expressed at low levels in resting endothelial cells and can be markedly upregulated in response to inflammatory mediators such as TNF-α, interleukin 1 and interferon-γ on a variety of cell types. ICAM-1 plays a role in the extravasation of leukocytes from the vasculature to inflamed tissue and activation of leukocytes in the inflamed tissue (Springer, 1990; Butcher, 1991; Dustin and Springer, 1991). Screening multiple first-generation phosphorothioate oligodeoxynucleotides targeting various regions of the human ICAM-1 transcript resulted in the identification of several oligonucleotides which were effective at inhibiting ICAM-1 expression in a variety of cell types (Chiang *et al.*, 1991; Bennett *et al.*, 1994). The most effective first-generation phosphorothioate oligodeoxynucleotides identified, ISIS 1939 and ISIS 2302, targeted specific sequences in the 3′- untranslated region of the human ICAM-1 mRNA. Both ISIS 1939 and ISIS 2302 inhibit ICAM-1 expression by an RNase H dependent mechanism of action (Bennett *et al.*, 1994) ISIS 2302 was selected for further additional studies.

ISIS 2302 selectively inhibits ICAM-1 expression in a variety of cell types (Bennett *et al.*, 1994; Miele *et al.*, 1994; Nestle *et al.*, 1994). Both sense and a variety of scrambled control oligonucleotides failed to inhibit ICAM-1 expression, including a 2-base mismatch control (Bennett *et al.*, 1994; Miele *et al.*, 1994; Nestle *et al.*, 1994). Treatment of endothelial cells with ISIS 2302 blocked adhesion of leukocytes, demonstrating that blocking expression of ICAM-1 will attenuate adhesion of leukocytes to activated endothelial cells (Bennett *et al.*, 1994). ISIS 2302 also blocked a one-way mixed lymphocyte reaction when the antigen-presenting cell was pretreated with ISIS 2302 to down-regulate ICAM-1 expression prior to exposure to the lymphocyte (T. Vickers *et al.*, unpublished data). Thus, ISIS 2302 is capable of blocking both leukocyte adhesion to activated endothelial cells and co-stimulatory signals to T lymphocytes, both activities being predicted based on previous studies with monoclonal antibodies to ICAM-1.

ISIS 2302 is selective for human ICAM-1 mRNA, limiting its application for *in vivo* pharmacology studies. To test the pharmacology of the human-specific antisense oligonucleotide we used experimental models in which immunocompromised mice contain human tissue xenografts. In one model we were able to demonstrate a role for ICAM-1 in the metastasis of human melanoma cells to the lung of mice (Miele *et al.*, 1994). A second study addressed the role of ICAM-1 in the production of cytotoxic dermatitis (lichen planus) in SCID mice containing human skin xenografts (Yan *et al.*, 1993; Murray *et al.*, 1994).

Upon engraftment of the human tissue, heterologous lymphocytes injected into the graft migrate into the epidermis (epidermaltropism) and produce a cytotoxic interaction between effector lymphocytes and epidermal cells (Christofidou-Solomidou *et al.*, 1997). Migration of the lymphocytes into the epidermis was correlated with expression of ICAM-1 in the epidermis. Systemic administration of ISIS 2302 inhibited ICAM-1 expression in the human graft, decreased the migration of lymphocytes into the epidermis and prevented subsequent lesion formation. A sense control oligodeoxynucleotide failed to attenuate the responses. These data demonstrate than an ICAM-1 antisense oligonucleotide administered systemically can attenuate an inflammatory response in the skin.

ISIS 3082 and ISIS 9125 are 20-base phosphorothioate oligodeoxynucleotides which hybridize to an analogous region in the 3′ untranslated region of murine and rat ICAM-1 mRNA, respectively. Similarly to ISIS 2302, ISIS 3082 and ISIS 9125 selectively inhibit ICAM-1 expression in mouse or rat cells by an RNase H dependent mechanism (Stepkowski *et al.*, 1994).

Previous studies have demonstrated that monoclonal antibodies to ICAM-1 prolong heterotopic cardiac allograft survival in mice (Isobe *et al.*, 1992). ISIS 3082 was tested in the same model to determine if an ICAM-1 antisense oligonucleotide would prolong cardiac allografts (Stepkowski *et al.*, 1994). Treatment of recipient C3H mice with ISIS 3082 for 7 days or 14 days by continuous intravenous infusion resulted in a dose-dependent prolongation of C57BL/10 cardiac allograft survival. Maximal effects occurred at 5–10 mg/kg per day. Treatment of recipient mice with 5mg/kg per day for 14 days increased cardiac allograft survival from 7.7 ± 1.4 days to 23.0 ± 7.5 days. Similar results were obtained with two additional strain combinations. Two control phosphorothioate oligodeoxynucleotides failed to prolong cardiac allograft survival. ISIS 3082 was either additive of synergistic with antilymphocyte serum, brequinar or rapamycin in prolonging cardiac allograft survival. Similar to previous reports using monoclonal antibodies to ICAM-1 and LFA-1 (Isobe et al., 1992), the combination of ISIS 3082 plus a monoclonal antibody to LFA-1 increased survival of the cardiac allograft to greater than 150 days. These results suggest that the combination of an LFA-1 monoclonal antibody and an inhibitor if ICAM-1 (either an antibody or antisense oligonucleotide) induces donor-specific transplantation tolerance. In the mouse model of cardiac allograft rejection, the combination of ISIS 3082 plus cyclosporin A attenuated the effect of each agent when given alone. This apparent antagonism between ISIS 3082 and cyclosporin A was unique to the mouse heterotopic heart model. ISIS 9125 (the rat-specific oligonucleotide) is synergistic with cyclosporin A in rat kidney and heart allograft models (S. Stepkowski *et al.*, unpublished data).

ISIS 3082 also prolonged survival of mouse islet cell allografts, demonstrating that the effects are not restricted to the heart (Katz et al., 1995). The rat ICAM-1 antisense oligonucleotide, ISIS 9125 prolongs survival of rat cardiac and kidney allografts in a dose-dependent manner (S. Stepkowski *et al.* unpublished data). The effects of the oligonucleotide were more pronounced in the kidney allograft model, which is consistent with the pharmacokinetics of phosphorothioate oligodeoxynucleotides in that the kidney is the major organ of disposition (Agrawal, Temsamani and Tang, 1991; Cossum *et al.*, 1993; Crooke *et al.*, 1996). Perfusion of the kidney allograft or treatment of the donor animal with ISIS 9125, but not the control oligodeoxynucleotide, prolonged survival of the allograft, suggesting that the effects of the oligonucleotide are on the donor tissue rather than the recipient.

Increased expression of ICAM-1 has been detected in both ulcerative colitis and Crohn's disease (Koizumi *et al.*, 1992; Schuermann *et al.*, 1993). The murine-specific ICAM-1 antisense oligonucleotide, ISIS 3082 was evaluated in a dextran sulfate model of colitis in mice (Bennett et al., 1997). Mice treated with dextran sulfate for 7 days exhibited increased ICAM-1 expression on endothelial cells in the submucosa, in lymphoid structures and on infiltrating leukocytes, demonstrating that ICAM-1 was expressed in inflamed colonic tissue. In normal tissue, the ICAM-1 oligonucleotide was localized in the lamina propria and to a lesser extent in epithelial cells in normal mice. By contrast, in animals with active colitis, epithelial cells accumulated significantly more of the oligonucleotide compared with normal animals (Bennett *et al.*, 1997). Treatment of mice with ISIS 3082 decreased ICAM-1 expression and leukocyte infiltration into the colon of dextran sulfate treated mice. ISIS 3082 was effective in preventing the development of colitis when administered prophylactically and also in attenuating existing colitis. The optimal dose for preventing development of colitis was 0.3–1.0 mg/kg per day, while approximately 10-fold higher concentrations were required to attenuate existing disease. Several control oligonucleotides also were evaluated in the model and found to have minimal effects.

Haller *et al.* (1996) used an ICAM-1 antisense oligonucleotide to decrease acute renal injury following ischaemia in rats. They identified a 20-base phosphorothioate oligodeoxynucleotide targeting the 3′ untranslated region of ICAM-1 mRNA. This oligonucleotide was shown to inhibit ICAM-1 expression in rat cells in a sequence specific manner. Using a cationic lipid formulation of the oligonucleotide infused into the femoral vein, they demonstrated decreased ICAM-1 protein expression following ischaemic injury and decreased leukocyte infiltrate (Haller *et al.*, 1996). The ICAM-1 antisense oligonucleotide also preserved renal function since blood urea nitrogen and serum creatine were reduced in the antisense oligonucleotide treated group 12–24 h after injury compared with saline or the control oligonucleotide treated groups. These data suggest that inhibition of ICAM-1 expression or function protects against ischaemia-reperfusion injury in kidney.

12.2.3 Toxicology

At least four distinct types of toxicity for phosphorothioate oligodeoxynucleotides are possible: sequence specific toxicity due to exaggerated pharmacology, sequence specific toxicity due to serendipitous hybridization to non-target RNA, sequence specific toxicity due to non-antisense effects, and sequence independent, non-antisense effects. There are no reported examples to date of toxicities which can be attributed to either exaggerated pharmacology or serendipitous inhibition of a non-targeted RNA. This may be due in part to selection of molecular targets that would not be expected to produce adverse effect due to inhibition of their expression, such as viral gene products, or targets for which there are redundant pathways in normal organisms. In addition, it should be kept in mind based upon pharmacokinetics that, in contrast to genetic knockouts, there would be a range of sensitivities of different organs and cell types within organs to the effects of oligonucleotides. Thus, the major toxicology concerns are class effects of phosphorothioate oligodeoxynucleotides due to interaction with non-nucleic acid targets (Henry *et al.*, 1997a; Levin *et al.*, 1997; Monteith *et al.*, 1997).

Studies in rodents and primates have revealed differential species sensitivities to the effects of phosphorothioate oligodeoxynucleotides. In rodents, activation of cells in the immune system appears to be the dose limiting toxicity, while in primates, acute effects on the complement system and anti-coagulant effects are of primary concern (Levin *et al.*, 1997). Initial studies in which phosphorothioate oligodeoxynucleotides were administered chronically to rodents revealed that the oligonucleotides produced polyclonal B cell activation and splenomegaly (Krieg *et al.*, 1989; Branda *et al.*, 1993; Pisetsky and Reich, 1993, 1994). In some studies this effect was attributed to a pharmacological effect of the oligonucleotide (Krieg *et al.*, 1989; Branda *et al.*, 1993); however, more recent studies suggest that this a common effect of all phosphorothioate oligodeoxynucleotides. A comparison of the effects of phosphorothioate oligodeoxynucleotides on murine and primate B lymphocytes demonstrates that rodent lymphocytes are more sensitive to the oligonucleotides, and the magnitude of response is more dramatic in mice compared with either human or cynomolgus monkey lymphocytes (Liang *et al.*, 1996; F. Bennett *et al.*, unpublished data). Some oligonucleotide sequences are particularly effective at activating B lymphocytes, which appears to be related to the presence of CpG motifs flanked by two purines on the 5′ side and pyrimidines on the 3′ side of the CpG (Krieg *et al.*, 1995). In addition to splenomegaly, mononuclear cell infiltrates in liver and kidney are observed following chronic administration of high doses of oligonucleotides (Dean *et al.*, 1996b; Bennett *et al.*, 1997; Levin *et al*, 1997). Other immune cell types which appear to be sensitive to phosphorothioate oligodeoxynucleotides include monocytes, neutrophils and NK cells (Kuramoto *et al.*, 1992; Stacey, Sweet and Hume, 1996; Ballas, Rasmussen and Krieg, 1996; Benimetskaya *et al.*, 1997; Boggs *et al.*, 1997).

The mechanism(s) by which oligonucleotides activate cells of the immune system are not understood fully. Data strongly suggest that oligonucleotides are activating signal transduction pathways in B lymphocytes. Initial studies by Krieg *et al.* (1995) suggested that internalization of the oligonucleotide was required for activation of murine lymphocytes. However, Liang *et al.* (1996) demonstrated that internalization of oligonucleotides was not required for activation of human B lymphocytes. CpG containing oligonucleotides attenuate anti-IgM-induced apoptosis in murine B cell lymphomas through activation of c-*myc* and *bcl*-X_L (Yi *et al.*, 1996b). Activation of murine lymphocytes is inhibited by antioxidants, suggesting that oligonucleotides are activating a reactive oxygen intermediate-dependent signal transduction pathway (Yi *et al.*, 1996a). We have examined the expression of acute activation markers on human leukocytes following treatment with phosphorothioate oligodeoxynucleotides. Phosphorothioate oligodeoxynucleotides increase expression of CD69 within 4 h on human B lymphocytes and monocytes, but not T lymphocytes, in a concentration dependent manner. Thus, activation of human leukocytes occurs rapidly after exposure to the oligonucleotide. The response of human neutrophils to phosphorothioate oligodeoxynucleotides is even more rapid: increased CD11b (MAC-1) expression and L-selectin shedding is detected within 5 min after adding to the cells (F. Bennett, unpublished data). These results suggest that this effect is due to direct interaction of the oligonucleotide with proteins on the surface of neutrophils capable of signalling neutrophils. MAC-1 would be a likely candidate (Benimetskaya *et al.*, 1997). In the presence of serum or in whole blood, the sensitivity of neutrophils is shifted by at least 100-fold.

Changes in blood pressure have been noted in monkeys given bolus intravenous injection of phosphorothioate oligodeoxynucleotides at dose levels of 5 mg/kg or greater (Cornish *et al.*, 1993; Galbraith *et al.*, 1994; Henry *et al.*, 1997b).

Haemodynamic changes have been characterized as a transient increase followed by a prolonged decrease in blood pressure. The changes in blood pressure are preceded by complement activation. Activation of complement appears to be related to peak plasma concentration, which can be avoided by infusion of the oligonucleotide over prolonged periods of time. The plasma concentration required to trigger complement activation is variable from animal to animal, but there does appear to a minimum peak threshold concentration of 50 μg/ml required to activate complement (Henry *et al.*, 1997b; Levin *et al.*, 1997). The mechanism(s) by which phosphorothioate oligodeoxynucleotides activate complement has not been well characterized. Preliminary data suggest that oligonucleotides bind to Factor H, subsequently inducing a loss of circulating Factor H. Since Factor H plays a key role in regulating complement activation, loss in circulating Factor H may, in part, mediate complement activation (Henry *et al.*, 1997b).

Phosphorothioate oligodeoxynucleotides also produce transient anti-coagulant effects in primates, which can be monitored by changes in activated partial thromboplastin time (aPTT). In contrast to complement activation, prolongation in aPTT is linear with respect to plasma concentration and as plasma concentrations of the drug decrease then changes in aPTT return to normal (Levin *et al.*, 1997). As with complement activation, risks associated with anti-coagulant effects can be reduced by prolonged infusion of the drug, keeping peak plasma concentration below threshold values.

Although phosphorothioate oligodeoxynucleotides do produce untoward effects in rodents and primates, these effects occur at dose levels greater than required to produce pharmacological activity. Therefore, it is anticipated that these drugs will be tolerated when administered to humans at pharmacologically relevant doses. Furthermore, by keeping peak plasma concentrations below threshold values, either by decreasing rates of absorption or prolonging rates of infusion, risks to patients can be minimized.

12.3 CLINICAL EXPERIENCE WITH ANTISENSE OLIGONUCLEOTIDES

To date twelve different oligonucleotides have entered clinical trials, with eight compounds still in active clinical development (Table 12.1). Similar to any other class of drugs, it can be expected that there will be failures in the clinic due to a variety of reasons such as lack of efficacy, marketing considerations, toxicity, etc. It is hoped that because of the generic pharmacokinetics and chemical class-specific toxicity that the failure rates for antisense oligonucleotides will be lower than other classes of agents. However, this remains to be seen.

Of the drugs in the clinic, the most advanced product is ISIS 2922 (Formivirsen) which is in Phase II/III studies in patients with CMV retinitis. ISIS 2922 is administered by direct intravitreal injection at doses ranging from 75 μg to 450 μg per eye. In a phase I study in patients which had failed intravenous anti-CMV therapy, the 450 μg dose was not well tolerated. Anterior chamber inflammation was observed in approximately one third of the patients in the other dose groups (Hutcherson *et al.*, 1995; Palestine *et al.*, 1995). Topical steroids controlled this inflammation in 8 out of 10 patients. ISIS 2922 produced a dose dependent antiviral effect with 0/2 eyes responding to the 75 μg dose, 2/4 eyes responding to the 150 μg dose, 6/10 eyes responding to the 300 μg dose, and 1/1 eye responding to the 450 μg dose. Time to progression in patients ranged from 3 weeks to 40 weeks with a median time to progression of 10 weeks. These results suggest that ISIS 2922 may be administered safely to patients at the 300 μg or lower dose and that the oligonucleotide exhibits anti-CMV activity. Based upon these data, pivotal studies were initiated for ISIS 2922 and, if successful, ISIS 2922 may be the first marketed antisense product.

Phase I/II clinical studies were initiated for GEM 91 in the United States and France (Kilkuskie and Field, 1997). The study performed in the United States was a randomized double-blind placebo controlled dose-escalating study in which GEM 91 was administered as a continuous intravenous infusion for two weeks, while in the French study GEM 91 was given as a 2 h intravenous infusion every other day for 28 days. Dose levels up to 4.4 mg/kg per day were achieved in the continuous infusion trials while dose levels of 3.0 mg/kg per day were reported for the intermittent infusion trial. Plasma half-lives for GEM 91 were biphasic with mean half-lives of 0.18 h and 26.7 h (Kilkuskie and Field, 1997; Zhang *et al*, 1995b). Hybridon recently announced the termination of clinical studies with GEM 91 based on lack of efficacy as measured by viral burden and the development of thrombocytopenia in some of the patients.

GEM 132 is a second generation chimeric molecule targeting the HCMV UL36 gene product. GEM 132 is a 20-mer oligonucleotide containing two 2′-*O*-methyl nucleosides on the 5′ end of the molecule and four 2′-*O*-methyl nucleotides on the 3′ end with the centre 14 residues being oligodeoxynucleotides (Kilkuskie and Field, 1997). The 2′-*O*-methyl residues confer increased hybridization affinity and increased nuclease resistance, while the centre oligodeoxynucleotide residues support RNase H activity. GEM 132 is being evaluated in CMV retinitis patients as both an intravenous infusion and as a direct intravitreal injection. In healthy volunteers, single 2 h infusions of GEM 132 were administered at doses ranging from 0.125 mg/kg to 0.5 mg/kg. Similar to phosphorothioate oligodeoxynucleotides, the plasma pharmacokinetics of GEM 132 were nonlinear with respect to dose. Single doses up to 0.5 mg/kg GEM 132 were tolerated well in normal volunteers, with headache being the most frequently reported side effect (Guinot *et al.*, 1997).

ISIS 2302, which targets human ICAM-1, is being developed jointly by Isis Pharmaceuticals, Inc. and Boehringer Ingelheim for the treatment of a variety of inflammatory disorders. Safety and pharmacokinetics of ISIS 2302 was established in a Phase I study performed at Guy's Hospital in normal volunteers (Glover *et al.*, 1997). Volunteers were either infused over a 2 h period with escalating single doses, or were given multiple doses of ISIS 2302 or saline in a double-blind trial. Brief dose-dependent increases in aPTT were seen at the time of peak plasma concentration and clinically insignificant increases in C3a were seen after repeated 2.0 mg/kg doses. C5a, blood pressure and pulse were unaffected by administration of ISIS 2302. No other adverse events or laboratory abnormalities related to the administration of the drug were noted. The Cmax was linearly related to dose and occurred at the end of infusion. Plasma half-life was approximately 53 min. Nonlinear changes in AUC and volume of distribution were noted with increasing dose, suggesting that oligonucleotide disposition might have a saturable component. These data suggest that ISIS 2302 was well tolerated in normal volunteers and that the pharmacokinetics in humans are similar to those observed in non-human primates and rodents.

Small Phase IIa studies (20–40 patients in each trial) have been initiated in rheumatoid arthritis, psoriasis, Crohn's disease, ulcerative colitis and renal transplant. With the exception of the psoriasis study, the trials are placebo controlled double-blind in which the drug is administered as a 2 h intravenous infusion. To date, only the Crohn's disease trial has been completed, by Dr. Bruce Yacyshyn at the University of Edmonton, in which patients were administered 0.5 mg/kg, 1.0 mg/kg and 2.0 mg/kg ISIS 2302 every other day for a total of 26 days. The response of the patients was not dose-dependent, probably due to the narrow dose range investigated and the small number of patients in the lower dose groups (3 each). Therefore, all ISIS 2302 treated patients were analysed as one group. Complete response, defined as a Crohn's disease activity index (CDAI) score less than 150, was observed in 7/15 patients treated with ISIS 2302 and 0/4 of the placebo patients (Yacyshyn *et al.*, 1997a,b). At the end of the study (6 months), 5 of the 7 patients were still in remission and 1 patient had a CDAI score of 156. During the treatment phase of the study, steroid doses were fixed; afterwards the physician was allowed to adjust steroid dose based upon symptoms. There was a statistically significant decrease in steroid use in patients treated with ISIS 2302 compared with placebo treated patients at the end of the study. Other than an expected increase in aPTT, and mild facial flushing at the end of infusion in one patient, the drug was tolerated well. Based upon these promising data, a large multicentre Phase IIb trial of ISIS 2302 in Crohn's disease has been initiated. Thus, ICAM-1 antisense oligonucleotides may have therapeutic utility for the treatment of Crohn's disease.

OL(1)p53 is a 20-mer phosphorothioate oligodeoxynucleotide complementary to a portion in exon 10 of the p53 mRNA (Bayever and Iversen, 1994; Bishop *et al.*, 1996). Preclinical studies with OL(1) p53 demonstrated that the p53 oligonucleotide inhibited proliferation of acute myelogenous leukaemia (AML) cells in cell culture. Correspondingly, OL(1) p53 was found to reduce the level of p53 in leukaemic cells, while a reverse sequence control failed to do so (Bishop *et al.*, 1996). A Phase I study was conducted at the University of Nebraska Medical Center in which OL(1) p53 was infused at doses ranging from 0.05 mg/kg per h to 0.25 mg/kg per h for 10 days into patients with haematological malignancies. There were no apparent toxicities that could be attributed directly to the oligonucleotide. Two patients experienced a transient increase in hepatic transaminase concurrent with administration of the drug. In contrast to observations made with ISIS 2302, 17–59% of intact drug was detected in urine in this group of patients. There was an inverse correlation between plasma concentrations of oligonucleotide and cumulative leukaemic growth of long term marrow cultures. However, this correlation was not observed clinically as there were no morphological complete responses. These results provide evidence that OL(1) p53 was tolerated in leukaemic patients.

Overexpression of *bcl*-2 is common in several cancers, in particular non-Hodgkin lymphoma, and may contribute to decreased sensitivity to chemotherapeutic agents (Reed *et al.*, 1990; Reed, 1995). An 18-mer phosphorothioate antisense oligodeoxynucleotide targeting the translational initiation codon of the *bcl*-2 gene was shown to inhibit the growth of lymphoma cells in SCID mice (Cotter *et al.*, 1994). Webb *et al.* (1997) conducted a Phase I clinical trial of this oligonucleotide (Genta 3139) at the Royal Marsden Hospital in London. Genta 3139 was administered as a daily subcutaneous infusion for 14 days to patients with *bcl*-2 positive non-Hodgkin lymphoma. The dose of the drug given ranged from 4.6 mg/m^2 to 73.6 mg/m^2. Other than local inflammation at the site of infusion, no treatment-related side effects were noted. In two patients, tomography scans revealed reductions in tumour size, with one complete response. In two additional patients the number of circulating lymphoma cells decreased during treatment. Reduced levels of *bcl*-2 protein expression in circulating lymphoma cells were detected in 2 out of 5 patients. These findings again demonstrate that phosphorothioate oligodeoxynucleotides can be administered safely to patients and also provide preliminary efficacy data with a *bcl*-2 antisense oligonucleotide.

Down-regulation of the C-*myb* transcription factor occurs during differentiation of haematopoietic cells, and C-*myb* protein expression appears to be necessary for the proliferation of these cells *in vitro* (Westin *et al.*, 1982). Inhibition of the colony-forming ability of normal bone progenitor cells has been demonstrated using phosphorothioate oligodeoxynucleotides targeting c-*myb* (Calabretta *et al.*, 1991). This oligonucleotide also reduced growth of primary AML and chronic myelogenous leukaemia (CML) cultures and proliferation of a T-cell leukaemia cell line (Anfossi, Gewirtz and Calabretta, 1989; Calabretta *et al.*, 1991). A phosphorothioate oligodeoxynucleotide version of the c-*myb* oligodeoxynucleotide inhibited the growth of K562 erythroleukaemia cells in SCID mice and prolonged survival of animals treated with the oligonucleotide (Ratajczak *et al.*, 1992). Based upon preclinical activity, an *ex vivo* bone marrow purging study was initiated at the University of Pennsylvania with eight patients. Human stem cells were incubated with the c-*myb* antisense oligonucleotide for 24 h prior to marrow cryopreservation. Patients were reinfused with their marrow following chemotherapy. One patient failed to engraft and 4/6 patients had normal leukocyte counts three months after engraftment. In follow-up, one patient was in haematological remission at 18 months, a second had 80% normal metaphases at 2 years, and a third exhibited a minor response. In a parallel study, 17 CML patients (four in chronic state and 13 with blast crisis) received systemic infusions of the C-*myb* phosphorothioate oligodeoxynucleotide continuously for seven days followed by one week drug-free period. No drug related toxicities were reported and one patient with CML in blast crisis, appeared to revert to chronic stage, surviving for 14 months (Gewirtz, 1997). These preliminary results suggest that the C-*myb* phosphorothioate oligodeoxynucleotide may have utility for the treatment of haematological malignancies.

Our understanding of the role of PKC in the pathogenesis of human malignancies continues to evolve. Numerous investigators have demonstrated that PKC-α is the most universally expressed member of the PKC isozyme family and therefore is a logical target for pharmacological intervention. PKC activity has been shown to be altered in certain human malignancies. For example, breast cancers have been demonstrated to have an increased level of PKC activity compared with normal breast tissue obtained from the same patient (O'Brian *et al.*, 1989). It was unknown from this study if the increase in PKC activity was the result of up-regulation of a specific isozyme or if the entire family of isozymes was up-regulated. PKC expression has been characterized in multiple human breast cancer cell lines. In another study it was reported that several oestrogen receptor (ER) negative cell lines express higher levels of PKC than ER positive cell lines, suggesting the presence of a negative correlation between ER expression and the level of PKC activity (Fabbro *et al.*, 1986). Total PKC activity (i.e., both membrane associated and total PKC activity) was found to be decreased in colonic adenomas and colonic carcinomas relative to healthy adjacent colonic tissue and colonic tissue obtained from normal volunteers (Kopp *et al.*, 1991).

Based on the available biological evidence implicating PKC in the pathogenesis of certain solid tumour types, and the broad spectrum of antitumour activity of ISIS 3521/CGP 64128A in the nude mouse xenograft implant model, a Phase I clinical trial was initiated by Novartis in collaboration with Isis Pharmaceuticals, Inc. A variety of tumours were evaluated in the Phase I trial, which was completed recently. In one trial, ISIS 3521/CGP 64128A was administered as a continuous 21-day infusion, then rested for 7 days. The cycle could be repeated if the treatments were tolerated and the tumour did not progress (Sikic *et al.*, 1997). In a preliminary report of the trial, one patient with colon cancer had stabilization of previously rising carcino embryonic antigen (CEA) for 4 months on treatment, and one ovarian cancer patient had stabilization of an enlarging abdominal mass for 4 months. There were no grade 3 or grade 4 toxicities reported. One patient displayed transient thrombocytopenia and one patient exhibited leukopenia. Based upon promising clinical results in Phase I studies and safety profile, Phase II studies of ISIS 3521/CGP 64128A have been initiated.

As discussed earlier, C-*raf* kinase plays a central role within the mitogen-activated protein kinase signal transduction pathway. The identification of mutations in *ras* gene products, which bind to C-*raf* kinase, resulting in transformation of cells, combined with the finding that C-*raf* kinase is over-expressed in some lung carcinomas, suggest that inhibition of C-*raf* kinase expression may be beneficial in the treatment of some cancers. Isis Pharmaceuticals, Inc. and Novartis have also initiated two Phase I studies for ISIS 5132/CGP 69846A targeting human C-*raf* kinase. The study designs are similar to the ISIS 3521/CG 64128A trials in which the drug is administered either as a continuous 21-day infusion or 2 h infusion three times weekly for 21 days. Based upon safety and efficacy, a Phase II trial for ISIS 5132/CGP 69846A has also been initiated.

The discovery of viral oncogenes in the mid 1960's was a major breakthrough in understanding the molecular origins of cancer, and led directly to the identification of the first human oncogene, *ras*, in 1982 (Chang *et al.*, 1982). Since then, the pharmaceutical industry has been searching for selective inhibitors of *ras* or *ras* function. Despite this intense effort, few selective inhibitors of *ras* gene products have entered the clinic. An antisense oligonucleotide targeting the Ha-*ras* gene product has begun clinical trials. ISIS 2503 targets the AUG translation initiation codon of the Ha-*ras* gene product (Monia *et al.*, 1992). Although the frequency of mutations in human cancers is significantly higher for the Ki-*ras* gene product, we have found that antisense oligonucleotides targeting Ha-*ras* gene exhibit broader antitumour effects when evaluated in human tumour xenograft models. In fact the Ha-*ras* antisense oligonucleotide was effective against human tumour xenografts known to contain a mutation in the Ki-*ras* gene. Isis Pharmaceuticals, Inc. has initiated a multicentre Phase I trial against a broad spectrum of cancers. Patients will receive ISIS 2503 as a continuous intravenous infusion for 2 weeks followed by a 1-week drug-free period. Patients will repeat the cycle as long as they tolerate the drug or tumours fail to respond to therapies. In a second planned study the drug will be administered in a more convenient schedule, i.e., a weekly 24 h infusion of ISIS 2503. Thus a first generation phosphorothioate oligodeoxynucleotide targeted against normal Ha-*ras* is the first selective inhibitor of *ras* function to enter clinical trials.

12.4 CONCLUSIONS

As is to be expected with first-generation technology, a number of undesirable properties have been identified for phosphorothioate oligodeoxynucleotides (Crooke and Bennett, 1996; Stein, 1996). Despite these limitations, it is possible to use phosphorothioate oligodeoxynucleotides to selectively inhibit the expression of a targeted RNA in cell culture and *in vivo*. The pharmacokinetics of phosphorothioate oligodeoxynucleotides are similar across species, and do not appear to exhibit major sequence specific differences. When dosed at high levels it is possible to identify toxicities in rodents and primates. However, at doses currently under evaluation in the clinic, phosphorothioate oligodeoxynucleotides have been tolerated well. In addition, there is evidence that phosphorothioate oligodeoxynucleotides may be providing clinical benefit to patients with viral infections, cancer and inflammatory diseases.

There are at least four areas where chemistry can add value to first generation drugs: to increase potency, decrease toxicity, alter pharmacokinetics and lower costs. Extensive medicinal chemistry efforts have been focused successfully on identifying improved antisense oligonucleotides, which address some of these issues. As an example, numerous modified oligonucleotides have been identified which have a higher affinity for target RNA than do phosphorothioate oligodeoxynucleotides (Nielsen *et al.*, 1991; Kawasaki *et al.*, 1993; Milligan, Matteucci and Martin, 1993; Wagner *et al.*, 1993; Altmann *et al.*, 1996; Matteucci, 1996). Oligonucleotide modifications have been identified which exhibit increased resistance to serum and cellular nucleases, enabling use of oligonucleotides that do not have phosphorothioate linkages. The tissue distribution of oligonucleotides may be altered with either chemical modifications or formulations (Agrawal *et al.*, 1995b; Zhang *et al.*, 1995a, 1996; Bennett *et al.*, 1996; Crooke *et al.*, 1996). Preliminary data suggest also that oral delivery of antisense oligonucleotides may be feasible (Agrawal *et al.*, 1995b). Finally, a number of modified oligonucleotides have been described which potentially exhibited fewer toxicities than first-generation phosphorothioate oligodeoxynucleotides (Zhao *et al.*, 1995; Henry *et al.*, 1996; Boggs *et al.*, 1997; Monteith *et al.*, 1997). However, as experience with these modified oligonucleotides is rather limited, it remains to be seen whether they will have a distinct toxicity profile.

In conclusion, first-generation phosphorothioate oligodeoxynucleotides have proved to be valuable pharmacological tools for the researcher, and should yield new therapies for the patient. Identification of improved second- and third-generation oligonucleotides should provide more valuable research tools and, more importantly, better therapeutics for patients.

REFERENCES

Agrawal, S., Ikeuchi, T., Sun, D., Sarin, P.S., Konopka, A., Maizel, J. and Zamecnik, P.C. (1989) Inhibition of human immunodeficiency virus in early infected and chronically infected cells by antisense oligodeoxynucleotides and their phosphorothioate analogues, *Proceedings of the National Academy of Sciences of the USA*, **86**, 7790–4.

Agrawal, S. and Tang, J.Y. (1992) GEM 91: an antisense oligonucleotide phosphorothioate as a therapeutic agent for AIDS, *Antisense Research and Development*, **2**, 261–6.

Agrawal, S., Temsamani, J., Galbraith, W. and Tang, J. (1995a) Pharmacokinetics of antisense oligodeoxynucleotides, *Clinical Pharmacokinetics*, **28**, 7–16.

Agrawal, S., Temsamani, J. and Tang, J.Y. (1991) Pharmacokinetics, biodistribution, and stability of oligodeoxynucleotide phosphorothioates in mice, *Proceedings of the National Academy of Sciences of the USA*, **88**, 7595–9.

Agrawal, S., Zhang, X., Lu, Z., Zhao, H., Tamburin, J.M., Yan, J., Cai, H., Diasio, R.B, Habus, I., Jiang, Z., Iyer, R.P., Yu, D. and Zhang, R. (1995b) Absorption, tissue distribution and *in vivo* stability in rats of a hybrid antisense oligonucleotide following oral administration, *Biochemical Pharmacology*, **50**, 571–6.

Altmann, K.-H., Dean, N.M., Fabbro, D., Freier, S.M., Geiger, T., Haner, R., Husken, D., Martin, P., Monia, B.P., Muller, M., Natt, F., Nicklin, P., Phillips, J., Pieles, U., Sasmor, H. and Moser, H.E. (1996) Second generation of antisense oligonucleotides: from nuclease resistance to biological efficacy in animals, *Chimia* **50**, 168–76.

Anderson, K.P., Fox, M.C., Brown-Driver, V., Martin, M.J. and Azad, R.F. (1996) Inhibition of human cytomegalovirus immediate-early gene expression by an antisense oligonucleotide complementary to immediate-early RNA, *Antimicrobial Agents and Chemotherapy*, **40**, 2004–11.

Anfossi, G., Gewirtz, A.M. and Calabretta, B. (1989) An oligomer complementary to c-myb induced mRNA inhibits proliferation of myeloid leukemia cell lines, *Proceedings of the National Academy of Sciences of the USA*, **86**, 3379–84.

Asaoka, Y., Nakamura, S., Yoshida, K. and Nishizuka, Y. (1992) Protein kinase C. calcium and phospholipid degradation, *Trends in Biochemical Sciences*, **17**, 414–17.

Azad, R.F., Driver, V.B., Tanaka, K., Crooke, R.M. and Anderson, K.P. (1993) Antiviral activity of a phosphorothioate oligonucleotide complementary to RNA of the human cytomegalovirus major immediate-early region, *Antimicrobial Agents and Chemotherapy*, **37**, 1945–54.

Ballas, Z.K., Rasmussen, W.L. and Krieg, A.M. (1996) Induction of NK activity in murine and human cells by CpG motifs in oligodeoxynucleotides and bacterial DNA, *Journal of Immunology*, **157**, 1840–5.

Barbour, S.E. and Dennis, E.A. (1993) Antisense inhibition of group II phospholipase A_2 expression blocks the production of prostaglandin E_2 by P388D_1 cells, *Journal of Biological Chemistry*, **268**, 21875–82.

Bayever, E. and Iversen, P. (1994) Oligonucleotides in the treatment of leukemia, *Hematology and Oncology*, **12**, 9–14.

Benimetskaya, L., Loike, J.D., Khaled, Z., Loike, G., Silverstein, S.C., Cao. L., El Khoury, J., Cai, T.-q. and Stein, C.A. (1997) Mac-1 (CD11b/CD18) is an oligodeoxynucleotide-binding protein, *Nature Medicine*, **3**, 414–20.

Bennett, C. F. (1997) Antisense oligonucleotides: Is the glass half full or half empty? *Biochemical Pharmacology*, **54**.

Bennett, C.F., Chiang, M.-Y., Chan, H., Shoemaker, J.E.E. and Mirabelli, C.K. (1992) Cationic lipids enhance cellular uptake and activity of phosphorothioate antisense oligonucleotides, *Molecular Pharmacology*, **41**, 1023–33.

Bennett, C.F., Condon, T., Grimm, S., Chan, H. and Chiang, M.-Y. (1994) Inhibition of endothelial cell-leukocyte adhesion molecule expression with antisense oligonucleotides, *Journal of Immunology*, **152**, 3530–40.

Bennett, C.F. and Crooke, S. T. (1994) Regulation of endothelial cell adhesion molecule expression with antisense oligonucleotides, *Advances in Pharmacology*., **28**, 1–43.

Bennett, C.F., Kornbrust, D., Henry, S., Stecker, K., Howard, R., Cooper, S., Dutson, S., Hall, W. and Jacoby, H.I. (1997) An ICAM-1 antisense oligonucleotide prevents and reverses dextran sulfate sodium-induced colitis in mice, *Journal of Pharmacology and Experimental Therapy*, **280**, 988–1000.

Bennett, C.F., Zuckerman, J.E., Kornbrust, D., Sasmor, H., Leeds, J.M. and Crooke, S.T. (1996) Pharmacokinetics in mice of a ^{3}H labeled phosphorothioate oligonucleotide formulated in the presence and absence of a cationic lipid, *Journal of Controlled Release*, **41**, 121–30.

Bijsterbosch, M.K., Manoharan, M., Rump, E.T., De Vrueh, R.L.A., van Veghel, R., Tivel, K.L., Biessen, E.A.L., Bennett, C.F., Cook, P.D. and van Berkel, T.J.C. (1997) *In vivo* fate of phosphorothioate antisense oligodeoxynucleotides: predominant uptake by scavenger receptors on endothelial cells, *Nucleic Acids Research*, **25**, 3290–6.

Bishop, M.R., Iversen, P.L., Bayever, E., Sharp, J.G., Greiner, T.C., Copple, B.L., Ruddon, R., Zon, G., Spinolo, J., Arneson, M., Armitage, J.O. and Kessinger, A. (1996) Phase I trial of an antisense oligonucleotide OL(1)p53 in hematologic malignancies, *Journal of Clinical Oncology*, **14**, 1320–6.

Boggs, R.T., McGraw, K., Condon, T., Flournoy, S., Villiet, P., Bennett, C.F. and Monia, B.P. (1997) Characterization and modulation of immune stimulation by modified oligonucleotides, *Antisense Nucleic Acid Drug Development*, **7**, 461–71.

Branch, A.D. (1996) A hitchhiker's guide to antisense and nonantisense biochemical pathways, *Hepatology*, **24**, 1517–29.

Branda, R.F., Moore, A.L., Mathews, L., McCormack, J.J. and Zon, G. (1993) Immune stimulation by an antisense oligomer complementary to the *rev* gene of HIV-1, *Biochemical Pharmacology*, **45**, 2037–43.

Butcher, E.C. (1991) Leukocyte-endothelial cell recognition: three (or more) steps to specificity and diversity, *Cell*, **67**, 1033–6.

Butler, M., Stecker, K., and Bennett, C.F. (1997) Cellular distribution of phosphorothioate oligodeoxynucleotides in normal rodent tissues. *Laboratory Investigation*, **77**, 379–88.

Calabretta, B., Sims, R.B., Valtieri, M., Caracciolo, D., Szczylik, C., Venturelli, D., Ratajczak, M., Beran, M. and Gewirtz, A.M. (1991) Normal and leukemic hematopoietic cells manifest differential sensitivity to inhibitory effects of c-*myb* antisense oligodeoxynucleotides; an *in vitro* study relevant to bone marrow purging, *Proceedings of the National Academy of Sciences of the USA*, **88**, 2351–5.

Chang, E.H., Furth, M.E., Scolnick, E.M. and Lowy, D.R. (1982) Tumorigenic transformation of mammalian cells induced by a normal human gene homologous to the oncogene of Harvey murine sarcoma virus, *Nature*, **297**, 479–83.

Chiang, M.-Y., Chan, H., Zounes, M.A., Freier, S.M., Lima, W.F. and Bennett, C.F. (1991) Antisense oligonucleotides inhibit intercellular adhesion molecule 1 expression by two distinct mechanisms, *Journal of Biological Chemistry*, **266**, 18162–71.

Christofidou-Solomidou, M., Albelda, S.M., Bennett, C.F. and Murphy, G.F. (1997) Experimental production and modulation of human cytotoxic dermatitis in human-murine chimeras. *American Journal of Pathology*, **150**, 631–9.

Cioffi, C.L., Garay, M., Johnson, J.F., McGraw, K., Boggs, R.T., Hreniuk, D. and Monia, B.P. (1996) Selective inhibition of a-*raf* and c-*raf* mRNA expression by antisense oligodeoxynucleotides in rat vascular smooth muscle cells: role of a-*raf* and c-*raf* in serum-induced proliferation, *Molecular Pharmacology*, **51**, 383–9.

Colige, A., Sokolov, B.P., Nugent, P., Baserga, R. and Prockop, D.J. (1993) Use of antisense oligonucleotide to inhibit expression of a mutated human procollagen gene (COL1A1) in transfected mouse 3T3 cells, *Biochemistry*, **32**, 7–11.

Cornish, K.G., Iversen, P., Smith, L., Arneson, M. and Bayever, E. (1993) Cardiovascular effects of a phosphorothioate oligonucleotide to p53 in the conscious rhesus monkey, *Pharmacological Communications*, **3**, 239–47.

Cossum, P.A., Sasmor, H., Dellinger, D., Truong, L., Cummins, L., Owens, S.R., Markham, P.M., Shea, J.P. and Crooke, S. (1993) Disposition of the ^{14}C-labeled phosphorothioate oligonucleotide ISIS 2105 after intravenous administration to rats, *Journal of Pharmacology and Experimental Therapy*, **267**, 1181–90.

Cossum, P.A., Truong, L., Owens, S.R., Markham, P.M. Shea, J.P. and Crooke, S.T. (1994) Pharmacokinetics of a 14C-labeled phosphorothioate oligonucleotide, ISIS 2105, after intradermal administration to rats, *Journal of Pharmacology and Experimental Therapy*, **269**, 89–94.

Cotter, F.E., Johnson, P., Hall, P., Pocock, C., Al Mahdi, N., Cowell, J.K. and Morgan, G. (1994) Antisense oligonucleotides suppress B-cell lymphoma growth in a SCID-hu mouse model, *Oncogene*, **9**, 3049–55.

Crooke, S.T. (1992) Therapeutic applications of oligonucleotides, *Annual Reviews of Pharmacology and Toxicology* **32**, 329–76.

Crooke, S.T. (1995) *Therapeutic Applications of Oligonucleotides*, R.G. Landes Company, Austin, TX.

Crooke, S.T. and Bennett, C.F. (1996) Progress in antisense oligonucleotide therapeutics, *Annual Reviews of Pharmocology and Toxicology*, **36**, 107–29.

Crooke, S.T., Graham, M.J., Zuckerman, J.E., Brooks, D., Conklin, B.S., Cummins, L.L., Greig, M.J., Guinosso, C.J., Kornbrust, D., Manoharan, M., Sasmor, H.M., Schleich, T., Tivel, K.L. and Griffey, R.H. (1996) Pharmacokinetic properties of several novel oligonucleotide analogs in mice, *Journal of Pharmacology and Experimental Therapy*, **277**, 923–37.

Daum, G., Eisenmann-Tappe, I., Fries, H.-W. and Troppmair, J. (1994) The ins and outs of *raf* kinases, *Trends in Biological Sciences*, **19**, 279–83.

De Clercq, E., Eckstein, F. and Merigan, T.C. (1969) Interferon induction increased through chemical modification of synthetic polyribonucleotide, *Science*, **165**, 1137–40.

Dean, N.M. and McKay, R. (1994) Inhibition of protein kinase C-alpha expression in mice after systemic administration of phosphorothioate antisense oligodeoxynucleotides, *Proceedings of the National Academy of Sciences of the USA*, **91**, 11762–6.

Dean, N.M. McKay, R., Condon, T.P. and Bennett, C.F. (1994) Inhibition of protein kinase C-α expression in human A549 cells by antisense oligonucleotides inhibits induction of intercellular adhesion molecule 1 (ICAM-1) mRNA by phorbol esters, *Journal of Biological Chemistry*, **269**, 16416–24.

Dean, N.M., McKay, R., Miraglia, L., Geiger, T., Muller, M., Fabbro, D. and Bennett, C.F. (1996a) Antisense oligonucleotides as inhibitors of signal transduction: development from research tools to therapeutic agents, *Biochemical Society Transactions*. **24**, 623–9.

Dean, N.M. McKay, R. Miraglia, L., Howard, R., Cooper, S., Giddings, J., Nicklin, P., Meister, L., Zeil, R., Geiger, T., Muller, M. and Fabbro, D. (1996b) Inhibition of growth of human tumor cell lines in nude mice by an antisense oligonucleotide inhibitor of PKC-alpha expression, *Cancer Research*, **56**, 3499–507.

Dekker, L.V. and Parker, P.J. (1994) Protein kinase C – a question of specificity, *Trends in Biochemical Sciences*, **19**, 73–7.

Dustin, M.L. and Springer, T.A. (1991) Role of lymphocyte adhesion receptors in transient interactions and cell locomotion, *Annual Reviews of Immunology*, **9**, 27–66.

Fabbro, D., Kung, W., Roos, W., Regazzi, R. and Eppenberger, U. (1986) Epidermal growth factor binding and protein kinase C activities in human breast cancer cell lines. *Cancer Research*, **46**, 2720–5.

Galbraith, W. M., Hobson, W.C., Giclas, P.C., Schechter, P.J. and Agrawal, S. (1994) Complement activation and hemodynamic changes following intravenous administration of phosphorothioate oligonucleotides in the monkey, *Antisense Research and Development*, **4**, 201–6.

Geary, R.S., Leeds, J.M., Fitchett, J., Burckin, T., Truong, L., Creek, M. and Levin, A.A. (1997b) Pharmacokinetics and metabolism in mice of a phosphorothioate oligonucleotide antisense inhibitor of C-*raf*-1 kinase expression. *Drug Metabolism Disposition*. In press.

Geary, R., Leeds, J., Henry, S.P., Monteith, D.K. and Levin, A.A. (1997a) Antisense oligonucleotide inhibitors for the treatment of cancer: 1) pharmacokinetic properties of phosphorothioate oligodeoxynucleotides, *Anti-Cancer Drug*, **12**, 383–93.

Geiger, T., Muller, M. Monia, B.P. and Fabbro, D. (1997) Antitumor activity of a C-*raf* antisense oligonucleotide in combination with standard chemotherapeutic agents against various human tumors transplanted subcutaneously into nude mice, *Clinical Cancer Research*, **3**, 1179–85.

Gewirtz, A.M. (1997) Developing oligonucleotide therapeutics for human leukemia, *Anti-cancer Drug Design*, **12**, 341–58.

Glover, J.M., Leeds, J.M., Mant, T.G.K., Kisner, D.L., Zuckerman, J., Levin, A.A. and Shanahan, W.R. (1997) Phase I safet and pharmacokinetic profile of an ICAM-1 antisense oligodeoxynucleotide (ISIS 2302), *Journal of Pharmacology and Experimental Therapy*. In press.

Graham, M.J., Crooke, S.T., Monteith, D.K., Cooper, S.R., M., L.-F.K., Martin, M.J. and Crooke, R.M. (1998) *In vivo* distribution and metabolism of a phosphorothioate oligonucleotide within rat liver after intravenous administration, *Journal of Pharmacology and Experimental Therapy*. In press.

Griffey, R.H., Monia, B.P., Cummins, L.L., Freier, S., Greig, M.J., Guinosso, C. J., Lesnik, E., Manalili, S.L., Mohan, V., Owens, S., Ross, B.R., Sasmor, H., Wancewicz, E., Weiler, K., Wheeler, P.D. and Cook, P.D. (1996) 2′-*O*-Aminopropyl ribonucleotides: a zwitterionic modification that enhances the exonuclease resistance and biological activity of antisense oligonucleotides, *Journal of Medicinal Chemistry*, **39**, 5100–9.

Guinot, P., Martin, R., Bonvoisin, B., Toneatt, C., Bourque, A., Cohen, A., Dvorchik, B. and Schechter, P. (1997) First phase I study of a new systemic anti-CMV antisense compound (GEM(r)132) in healthy male volunteers, 4th Conference Retroviruses and Opportunistic Infections, A742.

Haller, H., Dragun, D., Miethke, A., Park, J.K., Weis, A., Lippoldt, A., Grob, V. and Luft, F.C. (1996) Antisense oligonucleotides for ICAM-1 attenuate reperfusion injury and renal failure in the rat, *Kidney International*, **50**, 473–80.

Helene, C. and Toulme, J.-J. (1990) Specific regulation of gene expression by antisense, sense and antigene nucleic acids. *Biochimica et Biophysica Acta*, **1049**, 99–125.

Henry, S.P., Taylor, J., Midgley, L., Levin, A.A. and Kornbrust, D.J. (1997a) Evaluation of the toxicity profile of ISIS 2302, a phosphorothioate oligonucleotide, following repeated intravenous administration: 1) 4-week study in CD-1 mice, *Archives of Toxicology*. In press.

Henry, S.P., Zuckerman, J.E., Rojko, J., Hall, W.C., Harman, R.J., Kitchen, D. and Crooke, S.T. (1996) Toxicologic properties of several novel oligonucleotide analogs in mice *Anti-cancer Drug Design*, **12**, 1–14.

Henry, S.R., Giclas, P.C., Leeds, J., Pangburn, M., Auletta, C., Levin, A.A. and Kornbrust, D.J. (1997b) Activation of the alternative pathway of complement by a phosphorothioate oligonucleotide: potential mechanism of action, *Journal of Pharmacology and Experimental Therapy*, **281**, 810–16.

Hoke, G.D., Draper, K., Freier, S.M., Gonzalez, C., Driver, V.B., Zounes, M.C. and Ecker, D.J. (1991) Effects of phosphorothioate capping on antisense oligonucleotide stability, hybridization and antiviral efficacy versus herpes simplex virus infection, *Nucleic Acids Research*, **19**, 5743–8.

Howe, L.R., Leevers, S.J., Gomez, N., Nakielny, S., Cohen, P. and Marshall, C.J. (1992) Activation of the MAP kinase pathway by the protein kinase *raf*, Cell, **71**, 335–42.

Hutcherson, S.L., Palestine, A.G., Cantrill, H.L., Leiberman, R.M., Holland, G.N. and Anderson, K.P. (1995) Antisense oligonucleotide safety and efficacy for CMV retinitis in AIDS patients. *35th Interscience Conference on Antimicrobial Agents and Chemotherapy*, 204.

Isobe, M., Yagita, H., Okumura, K. and Ihara, A. (1992) Specific acceptance of cardiac allograft after treatment with antibodies to ICAM-1 and LFA-1, *Science*, **255**, 1125–7.

Kasid, U., Pfeifer, A., Brennan, T., Beckett, M., Weichselbaum, R.R., Dritschilo, A. and Mark, G.E. (1989) Effect of anisense c-*raf*-1 on tumorigenicity and radiation sensitivity of a human squamous carcinoma. *Science*, **243**, 1354–6.

Katz, S.M., Browne, B., Pham. T., Wang, M.E., Bennett, C.F., Stepkowski, S.M. and Kahan, B.D. (1995) Efficacy of ICAM-1 antisense oligonucleotide in pancreatic islet transplantation, *Transplantation Proceedings*, **27**, 3214.

Kawasaki, A.M., Casper, M.D., Freier, S.M., Lesnik, E.A., Zounes, M.C., Cummins, L.L., Gonzalez, C. and Cook, P.D. (1993) Uniformly modified 2′-deoxy-2′-fluoro phosphorothioate oligonucleotides as nuclease-resistant antisense compounds with high affinity and specificity for RNA targets, *Journal of Medicinal Chemistry*, **36**, 831–41.

Kilkuskie, R.E. and Field, A.K. (1997) Antisense inhibition of virus infections, *Advances in Pharmacology*, **40**, 437–83.

Koizumi, M., King, N., Lobb, R., Benjamin, C. and Podolsky, D.K. (1992) Expression of vascular adhesion molecules in inflammatory bowel disease, *Gastroenterology*, **103**, 840–7.

Kopp, R., Noelke, B., Sauter, G., Schildberg, F.W., Paumgartner, G. and Pfeiffer, A. (1991) Altered protein kinase C activity in biopsies of human colonic adenomas and carcinomas, *Cancer Research*, **51**, 205–10.

Koziolkiewicz, M., Wojcik, M., Kobylanska, A., Karwowski, B., Rebowska, B., Guga, P. and Stec, W.J. (1997) Stability of stereoregular oligo (nucleoside phosphorothioate)s in human plasma: diastereoselectivity of plasma 3′-exonuclease, *Antisense Nucleic Acid Drug Development*, **7**, 43–8.

Krieg, A., Gause, W.C., Gourley, M.F. and Steinberg, A.D. (1989) A role for endogenous retroviral sequences in the regulation of lymphocyte activation, *Journal of Immunology*, **143**, 2448–51.

Krieg, A.M., Yi, A.-K., Matson, S., Waldschmidt, T.J., Bishop, G.A., Teasdale, R., Koretzky, G.A. and Klinman, D.M. (1995) CpG motifs in bacterial DNA trigger direct B-cell activation *Nature*, **374**, 546–9.

Kumar, S., Tewary, H.K. and Iversen, P.L. (1995) Characterization of binding sites, extent of binding, and drug interactions of oligonucleotides with albumin, *Antisense Research and Development*, **5**, 131–9.

Kuramoto, E., Yano, O., Kimura, Y., Baba, M., Makino, T., Yamamoto, S., Yamamoto, T., Kataoka, T. and Tokunaga, T. (1992) Oligonucleotide sequences required for natural killer cell activation, *Japanese Journal of Cancer Research*, **83**, 1128–31.

Levesque, L., Dean, N.M., Sasmor, H. and Crooke, S.T. (1996) Antisense oligonucleotides targeting human protein kinase C-α inhibit phorbol ester-induced reduction of bradykinin-evoked calcium mobilization in A549 cells, *Molecular Pharmacology*, **51**, 209–16.

Levin, A.A., Monteith, D.K. Leeds, J.M., Nicklin, P.L., Geary, R.S., Butler, M., Templin, M.V. and Henry, S.P. (1997) Toxicity of oligodeoxynucleotide therapeutic agents, in *Antisense Research and Applications* (ed. S.T. Crooke), Springer-Verlag, Heidelberg.

Liang, H., Nishioka, Y., Reich, C.F., Pisetsky, D.S. and Lipsky, P.E. (1996) Activation of human B cells by phosphorothioate oligodeoxynucleotides, *Journal of Clinical Investigation*, **98**, 1119–29.

Liao, D.-F., Monia, B.P., Dean, N. and Berk, B.C. (1997) Protein kinase zeta mediates angiotensin II activation of ERK 1/2 in vascular smooth muscle cells, *Journal of Biological Chemistry*, **272**, 6146–50.

Lisziewicz, J., Sun, D., Metelev, V., Zamecnik, P., Gallo, R.C. and Agrawal, S. (1993) Long-term treatment of human immunodeficiency virus-infected cells with antisense oligonucleotide phosphorothioates, *Proceedings of the National Academy of Sciences of the USA*, **90**, 3860–4.

Lisziewicz, J., Sun, D., Weichold, F.F., Thierry, A.R., Lusso, P., Tang, J., Gallo, R.C. and Agrawal, S. (1994) Antisense oligodeoxynucleotides phosphorothioate complementary to Gag mRNA blocks replication of human immunodeficiency virus type 1 in human peripheral blood cells, *Proceedings of the National Academy of Sciences of the USA*, **91**, 7942–6.

Matteucci, M. (1996) Structural modifications toward improved antisense oligonucleotides. *Perspectives in Drug Discovery and Design*, **4**, 1–16.

McKay, R.M., Cummins, L.L., Graham, M.J., Lesnick, E.A., Owens, S.R., Winniman, M. and Dean, N.M. (1996) Enhanced activity of an antisense oligonucleotide targeting murine PKC-alpha by the incorporation of 2′-*O*-propyl modifications, *Nucleic Acid Research*, **24**, 411–17.

Miele, M.E., Bennett, C.F., Miller, B.E. and Welch, D.R. (1994) Enhanced metastatic ability of TNF-α-treated malignant melanoma cells is reduced by intercellular adhesion molecule-1 (ICAM-1, CD54) antisense oligonucleotides, *Experimental Cell Research*, **214**, 231–41.

Milligan, J.F., Matteucci, M.D. and Martin, J.C. (1993) Current concepts in antisense drug design, *Journal of Medicinal Chemistry*, **36**, 1923–37.

Monia, B., Johnston, J.F., Sasmor, H. and Cummins, L.L. (1996a) Nuclease resistance and antisense activity of modified oligonucleotides targeted to Ha-*ras*, *Journal of Biological Chemistry*, **24**, 14533–40.

Monia, B.P. (1997) Disruption of the MAP kinase signalling pathway using antisense oligonucleotide inhibitors targeted to ras and raf kinase, in *Applied Antisense Oligonucleotide Technology* (eds C. Stein and A. Krieg), Wiley, New York.

Monia, B.P., Johnston, J.F., Ecker, D.J., Zounes, M., Lima, W.F. and Freier, S.M. (1992) Selective inhibition of mutant Ha-*ras* mRNA expression by antisense oligonucleotides, *Journal of Biological Chemistry*, **267**, 19954–62.

Monia, B.P., Johnston, J.F., Geiger, T., Muller, M. and Fabbro, D. (1996b) Antitumor activity of a phosphorothioate oligodeoxynucleotide targeted against C-*raf* Kinase, *Nature Medicine*, **2**, 668–75.

Monia, B.P., Lesnik, E.A., Gonzalez, C., Lima, W.F., McGee, D., Guinosso, C.J., Kawasaki, A.M., Cook, P.D. and Freier, S.M. (1993) Evaluation of 2′ modified oligonucleotides containing deoxy gaps as antisense inhibitors of gene expression, *Journal of Biological Chemistry*, **268**, 14514–22.

Monia, B.P., Sasmor, H., Johnston, J.F., Freier, S.M., Lesnik, E.A, Muller, M., Geiger, T., Altmann, K.-H., Moser, H. and Fabbro, D. (1996c) Sequence-specific antitumor activity of a phosphorothioate oligodeoxyribonucleotide targeted to human c-*raf* kinase supports an antisense mechanism of action *in vivo*, *Proceedings of the National Academy of Sciences of the USA*, **93**, 15481–4.

Monteith, D.K., Henry, S.P., Howard, R.B., Flournoy, S., Levin, A.A., Bennett, C.F. and Crooke, S.T. (1997) Immune stimulation-A class effect of phosphorothioate oligonucleotides in rodents, *Anti-Cancer Drug Design*, **12**, 421–32.

Murray, A.G., Petzelbauer, P., Hughes, C.C., Costa, J., Askenase, P. and Pober, J.S. (1994) Human T-cell-mediated destruction of allogeneic dermal microvessels in a severe combined immunodeficient mouse. *Proceedings of the National Academy of Sciences of the USA*, **91**, 9146–50.

Nestle, F.O., Mitra, R.S., Bennett, C.F., Chan, H. and Nickoloff, B.J. (1994) Cationic lipid is not required for uptake and selective inhibitory activity of ICAM-1 phosphorothioate antisense oligonucleotides in keratinocytes, *Journal of Investigative Dermatology*, **103**, 569–75.

Nielsen, P.E., Egholm, M., Berg, R.H. and Buchardt, O. (1991) Sequence-selective recognition of DNA by strand displacement with a thymine-substituted polyamide, *Science*, **254**, 1497–500.

Nishizuka, Y. (1992) Intracellular signaling by hydrolysis of phospholipids and activation of protein kinase C, *Science*, **258**, 607–14.

O'Brian, C., Vogel, V.G., Singletary, S.E. and Ward, N.E. (1989) Elevated protein kinase C expression in human breast tumor biopsies relative to normal breast tissue, *Cancer Research*, **49**, 3215–17.

Palestine, A.G., Cantrill, H., Ai, E. and Lieberman, R. (1995) Treatment of cytomegalovirus (CMV) retinitis with ISIS 2922, *Investigative Opthalmology and Visual Science*, **36**, 858.

Pisetsky, D.S. and Reich, C. (1993) Stimulation of *in vitro* proliferation of murine lymphocytes by synthetic oligodeoxynucleotides, *Molecular Biology Reports*, **18**, 217–21.

Pisetsky, D.S. and Reich, C.F. (1994) Stimulation of murine lymphocyte proliferation by a phosphorothioate oligonucleotide with antisense activity for herpes simplex virus, *Life Sciences*, **54**, 101–7.

Plenat, F., Klein-Monhoven, N., Marie, B., Vignaud, J.-M. and Duprez, A. (1995) Cell and tissue distribution of synthetic oligonucleotides in healthy and tumor-bearing nude mice, *American Journal of Pathology*, **147**, 124–35.

Qian, M., Chen, S.-H., Von Hofe, E. and Gallo, J.M. (1997) Pharmacokinetics and tissue distribution of a DNA-methyltransferase antisense (MT-AS) oligonucleotide and its catabolites in tumor bearing nude mice, *Journal of Pharmacology and Experimental Therapy*, **282**, 663–70.

Rapp, U.R. (1991) Role of Raf-1 serine/threonine protein kinase in growth factor signal transduction, *Oncogene*, **6**, 495–600.

Rappaport, J., Hanss, B., Kopp, J.B., Copeland, T.D., Bruggeman, L.A., Coffman, T.M. and Klotman, P.E. (1995) Transport of phosphorothioate oligonucleotides in kidney: implications for molecular therapy, *Kidney International*, **47**, 1462–9.

Ratajczak, M.Z., Kant, J.A., Luger, S.M., Hijiya, N., Zhang, J., Zon, G. and Gewirtz, A.M. (1992) *In vivo* treatment of human leukemia in a *scid* mouse model with c-myb antisense oligodeoxynucleotides, *Proceedings of the National Academy of Sciences of the USA*, **89**, 11823–7.

Reed, J.C. (1995) Regulation of apoptosis by bcl-2 family proteins and its role in cancer and chemoresistance, *Current Opinions in Oncology*, **7**, 541–6.

Reed, J.C., Cuddy, M., Haldar, S., Croce, C., Nowell, P., Makover, D. and Bradley, K. (1990) *BCL2*-mediated tumorigenicity of a human T-lymphoid cell line: synergy with *MYC* and inhibition by *BCL2* antisense, *Proceedings of the National Academy of Sciences of the USA*, **87**, 3660–4.

Sands, H., Gorey-Feret, L.J., Cocuzza, A.J., Hobbs, F.W., Chidester, D. and Trainor, G.L. (1994) Biodistribution and metabolism of internally ^{3}H-labeled oligonucleotides. I. Comparison of a phosphodiester and a phosphorothioate, *Molecular Pharmacology*, **45**, 932–43.

Sanghvi, Y.S. and Cook, P.D. (1994) *Carbohydrate Modifications in Antisense Research*, ACS Symposium Series No. 580, American Chemical Society, Washington, DC.

Sawai, K., Takenori, M., Takakura, Y. and Hashida, M. (1995) Renal disposition characteristics of oligonucleotides modified at terminal linkages in perfused rat kidney, *Antisense Research and Development*, **5**, 279–87.

Schuermann, G.M., Aber-Bishop, A.E., Facer, P., Lee, J.C., Rampton, D.S., Dor, C.J. and Polak, J.M. (1993) Altered expression of cell adhesion molecules in uninvolved gut in inflammatory bowel disease, *Clinical and Experimental Immunology*, **94**, 341–7.

Schulte, T.W., Blagosklonny, M.V., Romanova, L., Mushinski, J.F., Monia, B.P., Johnston, J.F., Nguyen, P., Trepel, J. and Neckers, L.M. (1996) Destabilization of raf-1 by geldanamycin leads to disruption of the *raf*-1-MEK-mitogen-activated protein kinase signalling pathway, *Molecular and Cell Biology*, **16**, 5839–45.

Sharma, H.W. and Narayanan, R. (1995) The therapeutic potential of antisense oligonucleotides, *BioEssays*, **17**, 1055–63.

Shaw, J.-P., Kent, K., Bird, J., Fishback, J. and Froehler, B. (1991) Modified deoxyoligonucleotides stable to exonuclease degradation in serum, *Nucleic Acids Research*, **19**, 747–50.

Sikic, B.I., Yuen, A.R., Halsey, J., Fisher, G.A., Pribble, J.P., Smith, R.M. and Dorr, A. (1997) A phase I trial of an antisense oligonucleotide targeted to protein kinase C-alpha (ISIS 3521) delivered by 21-day continuous intrvenous infusion, *Proceedings of the American Society for Clinical Oncology*, **16**, 212a.

Soldatenkov, V.A., Dritschillo, A., Wang, F.-H., Olah, Z., Anderson, W.B. and Kasid, U. (1997) Inhibition of raf-1 protein kinase by antisense phosphorothioate oligodeoxyribonucleotide is associated with sensitization of human laryngeal squamous carcinoma cells to gamma radiation, *Cancer Journal Scientific American*, **3**, 13–20.

Springer, T.A. (1990) Adhesion receptors of the immune system, *Nature*, **346**, 425–34.

Stacey, K.J., Sweet, M.J. and Hume, D.A. (1996) Macrophages ingest and are activated by bacterial DNA, *Journal of Immunology*, **157**, 2116–22.

Stein, C.A. (1995) Does antisense exist? *Nature Medicine*, **1**, 1119–21.

Stein, C.A. (1996) Phosphorothioate antisense oligodeoxynucleotides: questions of specificity. *Trends in Biotechnology*, **14**, 147–9.

Stein, C.A. and Cheng, Y.-C. (1993) Antisense oligonucleotides as therapeutic agents – is the bullet really magical?, *Science*, **261**, 1004–12.

Stein, C.A. and Krieg, A.M. (1994) Problems in interpretation of data derived from *in vitro* and *in vivo* use of antisense oligodeoxynucleotides, *Antisense Research and Development*, **4**, 67–9.

Stein, C.A., Neckers, M., Nair, B.C., Mumbauer, S., Hoke, G. and Pal, R. (1991) Phosphorothioate oligodeoxycytidine interferes with binding of HIV-1 gp120 to CD4, *Journal of Acquired Immune Deficiency Syndrome*, **4**, 686–93.

Stein, C.A., Tonkinson, J.L. and Yakubov, L. (1991) Phosphorothioate oligodeoxynucleotides antisense inhibitors of gene expression, *Pharmacology and Therapy*, **52**, 365–84.

Stepkowski, S.M., Tu, Y., Condon, T.P. and Bennett, C.F. (1994) Blocking of heart allograft rejection by intercellular adhesion molecule-1 antisense oligonucleotides alone or in combination with other immunosuppressive modalities, *Journal of Immunology*, **153**, 5336–46.

Wagner, R.W. (1994) Gene inhibition using antisense oligodeoxynucleotides, *Nature*, **372**, 333–5.

Wagner, R.W., Matteucci, M.D., Lewis, J.G., Gutierrez, A.J. Moulds, C. and Froehler, B.C. (1993) Antisense gene inhibition by oligonucleotides containing C-5 propyne pyrimidines, *Science*, **260**, 1510–13.

Wang, H.-G., Miyashita, T., Takayama, S., Sato, T., Torigoe, T., Krajewski, S., Tanaka, S., Hovey, L., Troppmair, J., Rapp, R., Read, R.U. and Reed, J.C. (1994) Apoptosis regulation by interaction of bcl-2 protein and raf-1 kinase, *Oncogene*, **9**, 2751–6.

Webb, A., Cunningham, D., Cotter, F., Clarke, P.A., di Stefano, F., Corbo, M. and Dziewanowska, Z. (1997) BCL-2 antisense therapy in patients with non-Hodgkin lymphoma, *Lancet*, **349**, 1137–41.

Westin, E.H., Gallo, R.C., Arya, S.E., Eva, A., Souza, L.M., Baluda, M.A., Aaronson, S.A. and Wong-Staal, F. (1982) Differential expression of the AMV gene in human hematopoietic cells, *Proceedings of the National Academy of Sciences of the USA*, **79**, 2194–9.

Williams, N.G., Roberts, T.M. and Li, P. (1992) Both p21 *ras* and pp60 v-src are required, but neither alone is sufficient to activate the Raf-1 kinase, *Proceedings of the National Academy of Sciences of the USA*, **89**, 2922–6.

Yacyshyn B., Woloschuk, B., Yacyshyn, M.B., Martini, D., Doan, K., Tami, J., Bennett, F., Kisner, D. and Shanahan, W. (1997a) Efficacy and safety of ISIS 2302 (ICAM-1 antisense oligonucleotide) treatment of steroid-dependent Crohn's disease, *Gastroenterology*, **112**, A1123.

Yacyshyn, B.R., Bowen-Yacyshyn, M.B., Tami, J.A., Bennett, C.F., Kisner, D.L. and Shanahan, W.R. (1997b) A placebo-controlled trial of ISIS 2302 (ICAM-1 antisense oligonucleotide) in the treatment of steroid-dependent Crohn's disease.

Yan, H.-C., Juhasz, I., Pilewski, J.M., Murphy, G.F., Herlyn, M. and Albelda, S.M. (1993) Human/severe combined immunodeficient mouse chimeras: an experimental *in vivo* model system to study the regulation of human endothelial cel–leukocyte adhesion molecules, *Journal of Clinical Investigation*, **91**, 986–96.

Yazaki, T., Ahmad, S., Chahlavi, A., Zylber-Katz, E., Dean, N.M., Rabkin, S.D., Martuza, R.L. and Glazer, R.I. (1996) Treatment of glioblastoma U-87 by systemic administration of an antisense protein kinase C-α phosphorothioate oligodeoxynucleotide, *Molecular Pharmacology*, **50**, 236–42.

Yi, A.K., Klinman, D.M. Martin, T.L., Matson, S. and Krieg, A.M. (1996a) Rapid immune activation by CpG motifs in bacteria DNA. Systemic induction of IL-6 transcription through antioxidant-sensitive pathway, *Journal of Immunology*, **157**, 5394–402.

Yi, A.Y., Hornbeck, P., Lafrenz, D.E. and Krieg, A.M. (1996b) CpG DNA rescue of murine B lymphoma cells from anti-Igm-induced growth arrest and programmed cell death is associated with increased expression of c-myc and bcl-XL, *Journal of Immunology*, **157**, 4918–25.

Yuspa, S.H. (1994) The pathogenesis of squamous cell cancer: lessons learned from studies of skin carcinogenesis, *Cancer Research*, **54**, 1178–89.

Zhang, R., Iyer, R.P., Yu, D., Tan, W., Zhang, X., Lu, Z., Zhao, H. and Agrawal, S. (1996) Pharmacokinetics and tissue distribution of a chimeric oligodeoxynucleoside phosphorothioate in rats after intravenous administration, *Journal of Pharmacology and Experimental Therapy*, **278**, 971–9.

Zhang, R., Lu Z., Zhao, H., Zhang, X., Diasio, R.B., Habus, I., Jiang, Z., Iyer, R. P., Yu, D. and Agrawal, S. (1995a) *In vivo* stability, disposition and metabolism of a "hybrid' oligonucleotide phosphorothioate in rats, *Biochemical Pharmacology*, **50**, 545–56.

Zhang, R., Yan, J., Shahinian, H., Amin, G., Lu, Z., Liu, T., Saag, M.S., Jiang, Z., Temsamani, J., Martin, R.R., Schechter, P.J., Agrawal, S. and Diasio, R.B. (1995b) Pharmacokinetics of an anti-human immunodeficiency virus antisense oligodeoxynucleotide phosphorothioate (GEM 91) in HIV-infected subjects, Clinical Pharmacology and Therapy, **58**, 44–53.

Zhao, Q., Temsamani, J., Iadarola, P.L., Jiang, Z. and Agrawal, S. (1995) Effect of different chemically modified oligodeoxynucleotides on immune stimulation, *Biochemical Pharmacology*, **51**, 173–82.

Index